P9-CLE-869

the four

the
four

The Hidden DNA of
Amazon, Apple, Facebook, and Google

Scott Galloway

PORTFOLIO / PENGUIN

Delafield Public Library
Delafield, WI 53018
262-646-6230
www.delafieldlibrary.org

Portfolio/Penguin
An imprint of Penguin Random House LLC
375 Hudson Street
New York, New York 10014

Copyright © 2017 by Scott Galloway
Penguin supports copyright. Copyright fuels creativity, encourages diverse voices, promotes free speech, and creates a vibrant culture. Thank you for buying an authorized edition of this book and for complying with copyright laws by not reproducing, scanning, or distributing any part of it in any form without permission. You are supporting writers and allowing Penguin to continue to publish books for every reader.

Most Portfolio books are available at a discount when purchased in quantity for sales promotions or corporate use. Special editions, which include personalized covers, excerpts, and corporate imprints, can be created when purchased in large quantities. For more information, please call (212) 572-2232 or e-mail specialmarkets@penguinrandomhouse.com. Your local bookstore can also assist with discounted bulk purchases using the Penguin Random House corporate Business-to-Business program. For assistance in locating a participating retailer, e-mail B2B@penguinrandomhouse.com.

9780735213654 Hardcover
9780735213661 eBook
9780525533306 International

Printed in the United States of America
1   3   5   7   9   10   8   6   4   2

Illustrations by Kyle Scallon
Cover concept by Luidmilla Morozova
Book design by Daniel Lagin

3 0646 00218 6736

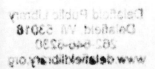

For Nolan & Alec
I look up, see the stars, and have questions.
I look down, see my boys, and have answers.

Contents

Contents

the four

Chapter 1

The Four

OVER THE LAST TWENTY YEARS, four technology giants have inspired more joy, connections, prosperity, and discovery than any entity in history. Along the way, Apple, Amazon, Facebook, and Google have created hundreds of thousands of high-paying jobs. The Four are responsible for an array of products and services that are entwined into the daily lives of billions of people. They've put a super-computer in your pocket, are bringing the internet into developing countries, and are mapping the Earth's land mass and oceans. The Four have generated unprecedented wealth ($2.3 trillion) that, via stock ownership, has helped millions of families across the planet build economic security. In sum, they make the world a better place.

The above is true, and this narrative is espoused, repeatedly, across thousands of media outlets and gatherings of the innovation class (universities, conferences, congressional hearings, boardrooms). However, consider another view.

The Four Horsemen

Imagine: a retailer that refuses to pay sales tax, treats its employees poorly, destroys hundreds of thousands of jobs, and yet is celebrated as a paragon of business innovation.

A computer company that withholds information about a domestic act of terrorism from federal investigators, with the support of a fan following that views the firm similar to a religion.

A social media firm that analyzes thousands of images of your children, activates your phone as a listening device, and sells this information to Fortune 500 companies.

An ad platform that commands, in some markets, a 90 percent share of the most lucrative sector in media, yet avoids anticompetitive regulation through aggressive litigation and lobbyists.

This narrative is also heard around the world, but in hushed tones. We know these companies aren't benevolent beings, yet we invite them into the most intimate areas of our lives. We willingly divulge personal updates, knowing they'll be used for profit. Our media elevate the executives running these companies to hero status—geniuses to be trusted and emulated. Our governments grant them special treatment regarding antitrust regulation, taxes, even labor laws. And investors bid their stocks up, providing near-infinite capital and firepower to attract the most talented people on the planet or crush adversaries.

So, are these entities the Four Horsemen of god, love, sex, and consumption? Or are they the Four Horsemen of the apocalypse? The answer is yes to both questions. I'll just call them the Four Horsemen.

How did these companies aggregate so much power? How can

an inanimate, for-profit enterprise become so deeply ingrained in our psyche that it reshapes the rules of what a company can do and be? What does unprecedented scale and influence mean for the future of business and the global economy? Are they destined, like other business titans before them, to be eclipsed by younger, sexier rivals? Or have they become so entrenched that nobody—individual, enterprise, government, or otherwise—stands a chance?

State of Affairs

This is where the Four stand at the time of this writing:

Amazon: Shopping for a Porsche Panamera Turbo S or a pair of Louboutin lace pumps is fun. Shopping for toothpaste and eco-friendly diapers is not. As the online retailer of choice for most Americans, and increasingly, the world, Amazon eases the pain of drudgery—getting the stuff you need to survive.[1,2] No great effort: no hunting, little gathering, just (one) clicking. Their formula: an unparalleled investment in last-mile infrastructure, made possible by an irrationally generous lender—retail investors who see the

MARKET CAPITALIZATION
AS OF APRIL 25, 2017

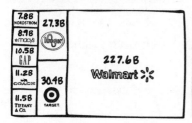
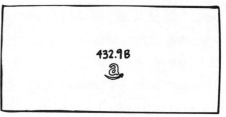

Yahoo! Finance. https://finance.yahoo.com/

most compelling, yet simple, story ever told in business: Earth's Biggest Store. The story is coupled with execution that rivals D-Day (minus the whole courage and sacrifice to save the world part). The result is a retailer worth more than Walmart, Target, Macy's, Kroger, Nordstrom, Tiffany & Co., Coach, Williams-Sonoma, Tesco, Ikea, Carrefour, and The Gap combined.[3]

As I write this, Jeff Bezos is the third wealthiest person in the world. He will soon be number one. The current gold and silver medalists, Bill Gates and Warren Buffet, are in great businesses (software and insurance), but neither sits on top of a company growing 20 percent plus each year, attacking multibillion-dollar sectors like befuddled prey.[4,5]

Apple: The Apple logo, which graces the most coveted laptops and mobile devices, is the global badge of wealth, education, and Western values. At its core, Apple fills two instinctual needs: to feel closer to God and be more attractive to the opposite sex. It mimics religion with its own belief system, objects of veneration, cult following, and Christ figure. It counts among its congregation the most important people in the world: the Innovation Class. By achieving a paradoxical goal in business—a low-cost product that sells for a premium price—Apple has become the most profitable company in history.[6] The equivalent is an auto firm with the margins of Ferrari and the production volumes of Toyota. In Q4 of 2016, Apple registered twice the net profits Amazon has produced, in total, since its founding twenty-three years ago.[7,8,9] Apple's cash on hand is nearly the GDP of Denmark.[10,11]

Facebook: As measured by adoption and usage, Facebook is the most successful thing in the history of humankind. There are 7.5 billion people in the world, and 1.2 billion people have a daily relationship with Facebook.[12,13] Facebook (#1), Facebook Messenger (#2), and Instagram (#8) are the most popular mobile apps in the United States.[14] The social network and its properties register fifty minutes of a user's typical day.[15] One of every six minutes online is spent on Facebook, and one in five minutes spent on mobile is on Facebook.[16]

Google: Google is a modern man's god. It's our source of knowledge—ever-present, aware of our deepest secrets, reassuring us where we are and where we need to go, answering questions from trivial to profound. No institution has the trust and credibility of Google: About one out of six queries posed to the search engine have never been asked before.[17] What rabbi, priest, scholar, or coach has so much gravitas that he or she is presented with that many questions never before asked of anybody? Who else inspires so many queries of the unknown from all corners of the world?

A subsidiary of Alphabet Inc., in 2016 Google earned $20 billion in profits, increased revenues 23 percent, and lowered cost to advertisers 11 percent—a massive blow to competitors. Google, unlike most products, ages in reverse, becoming more valuable with use.[18] It harnesses the power of 2 billion people, twenty-four hours a day, connected by their intentions (what you want) and decisions (what you chose), yielding a whole infinitely greater than the sum of its parts.[19] The insights into consumer behavior Google gleans from 3.5 billion queries each day make this horseman the executioner of traditional brands and media. Your new favorite brand is what Google returns to you in .0000005 second.

Show Me the Trillions

While billions of people derive significant value from these firms and their products, disturbingly few reap the economic benefits. General Motors created economic value of approximately $231,000 per employee (market cap/workforce).[20] This sounds impressive until you realize that Facebook has created an enterprise worth $20.5 *million* per employee . . . or almost a hundred times the value per employee of the organizational icon of the last century.[21,22] Imagine the economic output of a G-10 economy, generated by the population of Manhattan's Lower East Side.

The economic value accretion seems to be defying the law of big

RETURN ON HUMAN CAPITAL
2016

☑ NUMBER OF EMPLOYEES ☐ MARKET CAP PER EMPLOYEE

GM
215K
231 K

f
17,048
20.5M

Forbes, May, 2016. https://www.forbes.com/companies/general-motors/
Facebook, Inc. https://newsroom.fb.com/company-info/
Yahoo! Finance. https://finance.yahoo.com/

numbers and accelerating. In the last four years, April 1, 2013–April 1, 2017, the Four increased in value by approximately $1.3 trillion (GDP of Russia).[23,24]

Other tech companies, old and new, big and bigger, are losing relevance. Aging behemoths, including HP and IBM, barely warrant the attention of the Four. Thousands of start-ups fly by like gnats hardly worth swatting at. Any firm that begins to show the potential to bother the Four is acquired—at prices lesser companies can't

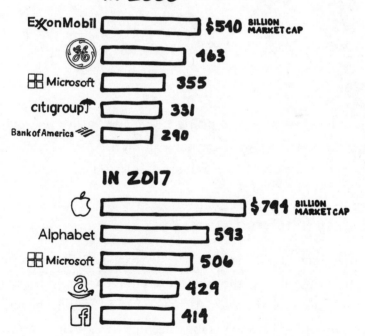

Taplin, Jonathan. "Is It Time to Break Up Google?" *The New York Times*.

WHERE PEOPLE START PRODUCT SEARCHES

2016

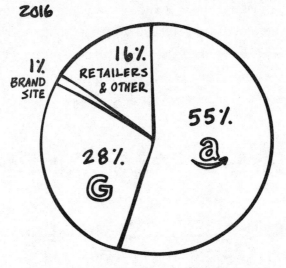

Soper, Spencer. "More Than 50% of Shoppers Turn First to Amazon in Product Search." Bloomberg.

imagine. (Facebook paid nearly $20 *billion* for five-year-old, fifty-employee instant messaging company WhatsApp.) Ultimately, the only competitors the Four face are . . . each other.

Safety in Hatred

Governments, laws, and smaller firms appear helpless to stop the march, regardless of the Four's impact on business, society, or the planet. However, there's safety in hatred. Specifically, the Four hate each other. They are now competing directly, as their respective sectors are running out of easy prey.

Google signaled the end of the brand era as consumers, armed with search, no longer need to defer to the brand, hurting Apple, who also finds itself competing with Amazon in music and film. Amazon is Google's largest customer, but it's also threatening Google in search—55 percent of people searching for a product start on Amazon (vs. 28 percent on search engines such as Google).[25] Apple and Amazon are running, full speed, into each other in front of us, on our TV screens and phones, as Google fights Apple to be the operating system of the product that defines our age, the smartphone.

Meanwhile, both Siri (Apple) and Alexa (Amazon) have entered the thunderdome, where two voices enter, and only one will leave. Among online advertisers, Facebook is now taking share from Google as it completes the great pivot from desktop to mobile. And the technology likely creating more wealth over the next decade, the cloud—a delivery of hosted services over the internet—features the Ali vs. Frazier battle of the tech age as Amazon and Google go head-to-head with their respective cloud offerings.

The Four are engaged in an epic race to become the operating system for our lives. The prize? A trillion-dollar-plus valuation, and power and influence greater than any entity in history.

So What?

To grasp the choices that ushered in the Four is to understand business and value creation in the digital age. In the first half of this book we'll examine each horseman and deconstruct their strategies and the lessons business leaders can draw from them.

In the second part of the book, we'll identify and set aside the mythology the Four allowed to flourish around the origins of their competitive advantage. Then we'll explore a new model for understanding how these companies exploit our basest instincts for growth and profitability, and show how the Four defend their markets with *analog moats*: real-world infrastructure designed to blunt attacks from potential competitors.

What are the horsemen's sins? How do they manipulate governments and competitors to steal IP? That's in chapter 8. Could there ever be a Fifth Horseman? In chapter 9, we'll evaluate the possible candidates, from Netflix to China's retail giant Alibaba, which dwarfs Amazon on many metrics. Do any of them have what it takes to develop a more dominant platform?

Finally, in chapter 10, we'll look at what professional attributes will help you thrive in the age of the Four. And in chapter 11, I'll talk about where the Four are taking us.

Alexa, Who Is Scott Galloway?

According to Alexa, "Scott Robert Galloway is an Australian professional football player who plays as a fullback for Central Coast Mariners in the A-League."

That bitch . . .

Anyway, while not a fullback, I've had a front-row seat to the Hunger Games of our age. I grew up in an upper-lower-middle-class household, raised by a superhero (single mother) who worked as a secretary. After college I spent two years at Morgan Stanley in a misguided attempt to be successful and impress women. Investment banking is an awful job, full stop. In addition, I don't have the skills—maturity, discipline, humility, respect for institutions—to work in a big firm (that is, someone else), so I became an entrepreneur.

After business school, I founded Prophet, a brand strategy firm that has grown to 400 people helping consumer brands mimic Apple. In 1997, I founded Red Envelope, a multichannel retailer that went public in 2002 and was slowly bled to death by Amazon. In 2010, I founded L2, a firm that benchmarks the social, search, mobile, and site performance of the world's largest consumer and retail brands. We use data to help Nike, Chanel, L'Oreal, P&G, and one in four of the world's one hundred largest consumer firms scale these four summits. In March 2017, L2 was acquired by Gartner (NYSE: IT).

Along the way, I've served on the boards of media companies (The New York Times Company, Dex Media, Advanstar)—all getting crushed by Google and Facebook. I also served on the board of Gateway, which sold three times more computers annually than Apple, at a fifth the margin—it didn't end well. Finally, I've also served on the boards of Urban Outfitters and Eddie Bauer, each trying to protect their turf from the great white shark of retail, Amazon.

However, my business card, which I don't have, reads "Professor of Marketing." In 2002, I joined the faculty of NYU's Stern School of Business, where I teach brand strategy and digital marketing and have taught over six thousand students. It's a privileged role for me, as I'm the first person, on either side of my family, to graduate from high school. I'm the product of big government, specifically the University of California, which decided, despite my being a remarkably unremarkable kid, to give me something remarkable: upward mobility through a world-class education.

The pillars of a business school education—which (remarkably) does accelerate students' average salaries from $70,000 (applicants) to $110,000 plus (graduates) in just twenty-four months—are Finance, Marketing, Operations, and Management. This curriculum takes up students' entire first year, and the skills learned serve them well the rest of their professional lives. The second year of business school is mostly a waste: elective (that is, irrelevant) courses that fulfill the teaching requirements of tenured faculty and enable the kids to drink beer and travel to gain fascinating (worthless) insight into "Doing Business in Chile," a real course at Stern that gives students credits toward graduation.

We require a second year so we can charge tuition of $110,000 vs. $50,000 to support a welfare program for the overeducated: tenure. If we (universities) are to continue raising tuition faster than inflation, and we will, we'll need to build a better foundation for the second year. I believe the business fundamentals of the first year need to be supplemented with similar insights into how these skills are applied in a modern economy. The pillars of the second year should be a study of the Four and the sectors they operate in (search, social, brand, and retail). To better understand these firms, the instincts they tap into, and their intersection between technology and stakeholder value is to gain insight into modern-day business, our world, and ourselves.

At the beginning and end of every course at NYU Stern, I tell my students the goal of the course is to provide them with an edge so they too can build economic security for themselves and their families. I wrote this book for the same reason. I hope the reader gains insight and a competitive edge in an economy where it's never been easier to be a billionaire, but it's never been harder to be a millionaire.

Chapter 2

Amazon

FORTY-FOUR PERCENT OF U.S. HOUSEHOLDS have a gun, and 52 percent have Amazon Prime.[1] Wealthy households are more likely to have Amazon Prime than a landline phone.[2] Half of all online growth and 21 percent of retail growth in the United States in 2016 could be attributed to Amazon.[3,4,5] When in a brick-and-mortar store, one in four consumers check user reviews on Amazon before purchasing.[6]

There are several good books, including Brad Stone's impressive *The Everything Store*, that tell the story of how a hedge fund analyst named Jeff Bezos drove cross-country from New York to Seattle with his wife and formulated his business plan for Amazon while on the road. Many who write about Amazon argue the firm's core assets are its operational capabilities, engineers, or brand. I, on the other hand, would argue that the real reasons Amazon is kicking the collective asses of its competition—and its likely ascent to a trillion dollars in value—are different.[7] Similar to the other Four, Amazon's

PERCENT OF AMERICAN HOUSEHOLDS 2016

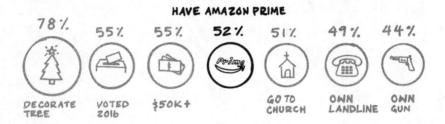

HAVE AMAZON PRIME

78%	55%	55%	52%	51%	49%	44%
DECORATE TREE	VOTED 2016	$50K+		GO TO CHURCH	OWN LANDLINE	OWN GUN

"Sizeable Gender Differences in Support of Bans on Assault Weapons, Large Clips." Pew Research Center.

ACTA, "The Vote Is In—78 Percent of U.S. Households Will Display Christmas Trees This Season: No Recount Necessary Says American Christmas Tree Association." ACTA.

"2016 November General Election Turnout Rates." United States Elections Project.

Stoffel, Brian. "The Average American Household's Income: Where Do You Stand?" *The Motley Fool.*

Green, Emma. "It's Hard to Go to Church." *The Atlantic.*

"Twenty Percent of U.S. Households View Landline Telephones as an Important Communication Choice." The Rand Corporation.

Tuttle, Brad. "Amazon Has Upper-Income Americans Wrapped Around Its Finger." *Time.*

rise rests on its appeal to our instincts. The other wind at its back is a simple, clear story that has enabled it to raise, and spend, staggering amounts of capital.

Hunters and Gatherers

Hunting and gathering, humanity's first and most successful adaptation, occupies more than 90 percent of human history.[8] By comparison, civilization is little more than a recent blip. It's less awful than it sounds: Paleolithic and Neolithic people spent just 10–20 hours a week hunting and gathering the food they needed to survive. The gatherers, in most cases women, were responsible for 80–90 percent of the effort and yield.[9] The hunters mostly provided extra protein.

This shouldn't be surprising. Men tend to be better at evaluating at a distance—where prey is first spotted. By comparison, women are typically better at taking stock of their immediate surroundings. Gatherers also needed to be more thoughtful about what they collected. While a tomato couldn't outrun her, the gatherer woman still needed to develop the skills needed to assess nuances such as ripeness, color, and shape for signs of edibility or disease. The hunter, by comparison, needed to act fast when the opportunity for a kill presented itself. There was no time for nuance, just speed and violence. Once the prey had been killed, the hunters needed to collect the merchandise and get home, pronto, as the fresh kill, and even themselves, were both attractive targets.[10]

Observe how women and men shop and you'll see that not much has changed. Women feel fabric, try on shoes with a dress, and ask to see things in different colors. Men see something that can sate their appetite, kill (buy) it, and get back to the cave as fast as possible.[11] For our distant ancestors, once the catch was safely back at the cave, the pile never seemed high enough. Famine threatened with every drought, snowstorm, or pestilence. So, over-collecting was a smart strategy: the downside of too much stuff was wasted effort. The downside of under-collecting was death from starvation.

Humans aren't alone in the compulsion to collect. For males of many animal species, collecting translates to sex. Consider male black wheatears, avian residents of dry and rocky regions in Eurasia and Africa. They hoard stones. The bigger the pile (the bigger the sales price of that loft in the Tribeca), the more females are interested in mating.[12] Like most neuroses, it starts with the best of intentions and then goes off the rails. Every year, there are scores of news stories of people being dug out from stuff that collapsed on them in the (dis)comfort of their home. That guy dug out by firemen from under forty-five years of newspapers isn't crazy—he was displaying his Darwinian fitness to anyone who dropped by.

Our Consuming Capitalist Selves

Instinct is a powerful chaperone, always watching and whispering in your ear, telling you what you *must* do to survive.

Instinct has a camera, but it's low resolution. It takes hundreds, if not thousands, of years to adapt. Take our affinity for salty, sugary, and fatty foods. It was a rational strategy in humanity's early days, as these ingredients were the most difficult to come by. Not so anymore. We have institutionalized the production of these food groups, like the Burger King Whopper or Wendy's Frosty, to easily meet our needs cost-effectively. Only, our instincts haven't caught up. By 2050, one in three Americans will likely have diabetes.[13]

Our hunger for more stuff hasn't adjusted to our limited closets and wallets, either. Many have a difficult time putting food on the table and affording the basics. Yet millions end up on anticholesterol drugs like Lipitor and with high-interest credit cards, because they can't maintain command over their powerful instinct to collect.

Instinct, coupled with a profit motive, makes for excess. And the worst economic system, except for all the rest—capitalism—is specifically designed to maximize that equation. Our economy and prosperity are largely predicated on others' consumption.

Fundamental to business is the notion that in a capitalist society the consumer reigns supreme, and consumption is the most noble of activities. Thus a country's place in the world is correlated with its level of consumer demand and production. After 9/11, President

George W. Bush's advice to a grieving nation was to "go down to Disney World in Florida, take your families and enjoy life the way we want it to be enjoyed."[14] Consumption has taken the place of shared sacrifice during times of war and economic malaise. The nation needs you to keep buying more stuff.

Few industries have created more wealth by tapping into our consuming selves than retail. Of the four hundred wealthiest people in the world (excluding those who inherited wealth or are in finance) more names on the list are in retail than even technology. Armancio Ortega, the scion of Zara, is the wealthiest man in Europe.[15] Number three, Bernard Arnault of LVMH, who may be thought of as the father of modern luxury, owns and operates 3,300-plus stores—more than Home Depot.[16,17] However, the well-publicized successes in retail, coupled with low barriers of entry and the dream of opening one's very own "shoppe," have created an industry that is over-stored and, like most industries, in a state of constant flux. Here is how "dynamic" the U.S. retail environment is:

- The top ten best-performing stocks of 1982 were Chrysler, Fay's Drug, Coleco, Winnebago, Telex, Mountain Medical, Pulte Home, Home Depot, CACI, and Digital Switch.[18] How many are still around today?
- The best-performing stock of the eighties? Circuit City (up 8,250 percent).[19] In case you don't remember, Circuit City was the now-bankrupt big-box store that sold TVs and other electronics, where *Service is State of the Art.* RIP.
- Of the ten biggest retailers in 1990, only *two* remain on the list in 2016.[20,21] Amazon, born in 1994, registered more revenue after twenty-two years in 2016 ($120 billion) than Walmart,

founded in 1962, did after thirty-five years in 1997 ($112 billion).[22,23]

In 2016, retail could largely be described as the crazy success of Amazon and the disaster that is the rest of the sector, with a few exceptions, such as Sephora, fast fashion, and Warby Parker. E-commerce firms die with a whimper, not a bang, because while brick-and-mortar retail has a face, e-commerce deaths are faceless and not as jarring. One day that website you regularly visited just isn't there—so you find some other site and never look back.

Dead man (retailer) walking begins with margin erosion—the cholesterol of retail—and ends with endless promotions and sales. You can buy a little time with sales, but the story almost always ends badly: holding an average of 12 percent more inventory in the December 2016 holiday season, retailers increased sales promotions from 34 percent to 52 percent.[24]

How did we get here? Let's take a brief walk down retail's memory lane. In the United States and Europe, there have been six major stages of retail evolution.[25]

The Corner Store

Retailing in the first half of the twentieth century was defined by the corner store. Proximity ruled the day. You walked to the store and carried what you could home, sometimes daily. Retail establishments were typically family run and played a key social role in the community, disseminating local news before radio and TV became dominant. Their competence was *Customer Relationship Management* (CRM), before that term was invented. Shop owners knew their clientele and would extend credit based upon your good name. Our love affair with retail and the nostalgia we feel when a legendary retailer files for bankruptcy (notice that it doesn't make the news when a venerable oil equipment leasing firm goes under) is a function of our historic affection for retail, which has been baked into our culture.

Department Stores

London's Harrods and Newcastle's Bainbridge's catered to a new market segment: emerging and affluent females who no longer felt bound by a chaperone. In London, the iconic Selfridges offered a hundred departments, restaurants, a roof garden, reading and writing rooms, reception areas for foreign visitors, a first-aid room, and knowledgeable floor people. Floor associates were trained and paid via a novel concept—the sales commission. The notion of differentiation through service, and of becoming the customer's temporary friend and shopping guide, broke new ground. It humanized large-scale retailing and redirected investing toward human capital at the store level. After Selfridges, these celebrations of architecture, lighting, fashion, consumerism, and community spread across Europe and the United States.

Department stores also reshaped the relationship between business and consumers. Traditionally, consumer businesses took on a paternal role, and told you what was best. The church/bank/store was in charge. You were supposed to feel fortunate to be blessed with the product of their collective wisdom. It was Harry Selfridge who coined the phrase "the customer is always right"—which at the time might have appeared weak and obsequious. In fact, it was profound and far-reaching: four of the five oldest surviving retailers are department stores: Bloomingdale's, Macy's, Lord & Taylor, and Brooks Brothers.[26]

Call of the Mall

As America barreled toward midcentury, the car and refrigerator meant we could drive farther to get more stuff we could store safely longer. Advances in distribution led to fewer visits, bigger stores, more selection, and lower prices. Department stores evolved into the mall. Also thanks to the automobile, the suburbs boomed. Developers responded by offering consumers a comfortable destination containing several different stores in one location connected by food courts and movie theaters. Malls became Main Street for suburbs that didn't have an obvious epicenter. (It has always baffled me how much pride people from Short Hills, New Jersey, have in their local mall. It's like owning a Quiznos franchise: I say keep it to yourself.) By 1987, half of U.S. retail sales were occurring in malls.[27]

But by 2016, business media was bemoaning the end of an American institution. Forty-four percent of the value of U.S. malls is in just a hundred places, and sales per square foot dropped 24 percent in the past decade.[28] A mall's health is more a reflection of the local economy than the format itself. Suburban blight has put many out of existence. However, many still thrive—particularly those that have a strong offering—a good mix of stores, parking, and proximity to the upper quartile of income-earning households.

The Big Box

1962 brought us the first American to orbit the Earth, the Cuban Missile Crisis, *The Beverly Hillbillies*—and Walmart, Target, and Kmart.

Big-box retail caused a dramatic shift in social norms and transformed the retail format. The notion of buying stuff in bulk and passing those savings onto consumers is not, on its own, revolutionary. More significant is that we, as a nation, decided to shift the consumer to the front of the line, in every way. At Home Depot, you could pick out your own lumber. At Best Buy you could shop every possible TV and take your choice home in your car.

Getting our stuff at the lowest possible price was now more important than any specific company, sector, or even the health of the broader community. The invisible hand began bitch-slapping small or inefficient retailers all over the United States and Europe. Mom-and-pop stores, previously a large part of community life, faced towering competition. The era also saw a new generation of retail infrastructure technology, including the first barcode scanner, installed in a Kroger in 1967.[29]

Until the sixties there were laws against retailers offering discounts for bulk purchases. Lawmakers correctly feared this would put thousands of local stores out of business. In addition, manufacturers' brands typically set the prices retailers were allowed to charge for their products. As a result, discounting was a limited and dull-edged weapon.

For various reasons, including falling margins and growing competition, those gloves came off in the sixties, and the great "Race to Zero" began. Today, on the homepage of hm.com one can find a long-sleeved ribbed mock turtleneck dress for just $9.99. For the same price you can also grab a men's textured fine-knit sweater. That's cheap, not only in today's dollars, but in 1962 dollars—a staggering achievement and a testament to a cut-throat race to the bottom.

As the shackles came off, the more-for-less big-box monsters created hundreds of billions in wealth. The next thirty years saw what was then the most valuable company and the world's wealthiest man, Sam Walton, emerge from this format, not to mention our collective view that the consumer now reigned supreme. People lament the job-destroying machine that is Amazon. But Walmart was the original gangster. The value proposition was clear and compelling: when you shop at Walmart, it's similar to getting a promotion—you get a better life, featuring Heineken instead of Budweiser beer, and Tide instead of Sun detergent.

Specialty Retail

Walmart was the great leveler. But most consumers don't want to be equal; they want to be *special*. And a sizable fraction of the consuming population will pay a premium for that attention. That fraction also tends to be the consumers with the most disposable income.

The march toward "more for less" created a vacuum for consumers looking for expertise and a social signal of something aspirational about their lives. Hence the rise of specialty retail, which enabled mostly affluent consumers to focus on an exclusive brand or product regardless of price. Thus Pottery Barn, Whole Foods, and Restoration Hardware.

A strong economy helped. This was the prosperous eighties, and young urban professionals found in these specialty stores their homes away from home—pleasure palaces where they could buy stuff for their homes and closets that better articulated just how cool and cultivated they were. You could find the right pork from an establishment that sold nothing but hams baked in honey, or get the perfect candle from a store that sold only candles (Illuminations), or look for some Linens and some Things. Many of these specialty retailers almost seamlessly transitioned into the era of e-commerce, as many had cut their teeth on direct-mail catalogs and were facile with data and fulfillment.

The retailer that truly defined the specialty retail era was The Gap. Rather than spending money on advertising, The Gap invested in store experience, becoming the first lifestyle brand. You felt cool shopping at The Gap, while buying a Pottery Barn couch gave a generation of Americans the sense that they had "arrived." Specialty retailers recognized that even shopping bags offered a self-expressive benefit—if you carried Williams-Sonoma, you were cool, enjoyed the finer things in life, and had a passion for cooking.

The E-Commerce Opportunity

Jeff Bezos happened more to retail than retail happened to Jeff Bezos. In each of the preceding eras of retail, there were brilliant people who tapped into a shift in demographics or taste and created billions of dollars in value. But Bezos saw a technological shift, then used it to reconstruct root and branch the entire world of retailing. E-commerce would be a shadow of itself, had Bezos not brought his vision and focus to the medium.

In the 1990s, e-commerce was a shitty, unrewarding business for almost every pure-play firm (it still is). The key to success in e-commerce wasn't execution but creating hype around a company's potential, and then selling it to some rich sucker before the house of cards caved in. The most current example is flash sale sites—sites that promised amazing deals but only at unspecified times. The press went wild. See a pattern? Hype does not equal sales.

Retail may have never been, on a risk-adjusted basis, a good business. But it was markedly less awful before Seattle's great white shark of retail showed up and began eating everything. Over the last decade, the market capitalization of the retail icons of the twentieth century—from Macy's to JCPenney's—has ranged from awful to

FLASH SALE SITES' INDUSTRY REVENUE

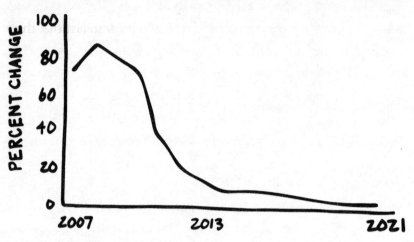

Lindsey, Kelsey. "Why the Flash Sale Boom May Be Over—And What's Next." RetailDIVE.

disastrous. There is a finite amount of capital invested in each sector, and Amazon's vision and execution has soaked up the preponderance of that investment. The result is a once-populous sector that is being ravaged and depopulated by a single player.

Because we live in a culture of consumption, the natural trajectory of retail is up. So, when the planets align and a new concept works, it can scale rapidly and create tremendous value for consumers and shareholders. Walmart really did give people access to a better, or at least more material, life. And you really can feel better about yourself wearing Zara Silver Patent Platform Oxfords and making juice in a Breville Juice Fountain from Williams-Sonoma.

The difference this time is that this value has been created with unprecedented speed by a *single* company, because, being virtual, Amazon can scale to hundreds of millions of customers, and scale across almost every retail industry, without the traditional drag of having to build brick-and-mortar stores and hire thousands of employees. On Amazon, Bezos realized, every page can be a store and every customer a salesperson. And the company could grow so fast that there wouldn't be any corners left for competitors to carve out a niche.

The Soon-to-Be Wealthiest Man on the Planet

In the first dot-com boom, Jeff Bezos was just another Wall Street escapee with a computer science degree who'd become enamored with the promise of e-commerce. But his vision and maniacal focus would set him head and shoulders above the rest. For his online shopfront, launched in Seattle in 1994, Bezos chose the name "Amazon" as an indicator of the scale of the flow of merchandise he envisioned. However, another name he considered (he still owns the URL) was more appropriate: *relentless.com*.[30]

When Bezos started Amazon, online shopping couldn't serve true gatherers because the limited web technology (lame experience) had the nuance and detail of a Lada, the Russian auto brand—ugly and underpowered. Brands are two things: promise and performance. The brand "internet" during the nineties and into the noughts was half that.

E-commerce in 1995 needed to be prey you recognized easily and could kill and take back to the cave with little loss of value or risk that you accidently brought a plant back that would poison the clan. Bezos decided this animal was . . . books.

Easy to recognize, kill, and digest. Books stacked in a warehouse, with a "look inside" preview. The prey has already been killed and stacked up for you. An industry—book reviewing—emerged to identify what books were worth eating/reading, bypassing the diligence of curation offered by a store. Bezos realized reviews could do the hard work of retailing for him. Amazon could call on the internet's less lame attributes: selection and distribution. So, no nuance like well-lit storefronts, a door chime, and friendly salespeople. Instead, he leased a warehouse near Seattle airport and filled it in a way that robots could maneuver easily.

In the early days, Amazon focused on books and hunters—people on a mission, looking for a specific product. As the years passed, broadband began to offer shades of nuance, and gatherers showed up, willing to browse, weigh options, and take their time. Bezos knew he could migrate to things people weren't used to buying online yet, like CDs and DVDs. Foreshadowing Amazon's threat to all things good in our society, Susan Boyle's CD *I Dreamed a Dream* set sales records on the platform.

To outrun competitors and reinforce the core value of selection, Amazon introduced Amazon Marketplace, letting third parties fill in the long tail. Sellers got access to the world's largest e-commerce platform and customer base, and Amazon was able to balloon its offerings without the expense of additional inventory.

Amazon Marketplace now accounts for $40 billion, or 40 percent, of Amazon's sales.[31] Sellers, content with the massive customer flow, feel no compulsion to invest in retail channels of their own. Meanwhile, Amazon gets the data and can enter any business (begin selling products themselves) the moment a category becomes attractive. So, Amazon, should it choose, can begin offering directly *"Old Asian Man Wall Decals," "Nicolas Cage Pillowcases,"* and *"55-Gallon Drums of Lube."*

Amazon appeals to our hunter-gatherer instinct to collect more stuff with minimum effort. We have serious mojo for stuff, as survival went to the caveman who had the most twigs, had the right rocks to crack stuff open with, and got the most colorful mud to draw images on walls so his descendants knew when to plant crops, or what dangerous animals to avoid.

The need for stuff is real: stuff keeps us warm and safe. It allows us to store and prepare food. It helps us attract mates and care for our offspring. And easy stuff is the best stuff, because it consumes less energy and gives you time to do other important things.

Without capital-hungry stores, Bezos could invest in automated warehouses. Scale is power, and Amazon was able to offer prices no brick-and-mortar retailer could afford. He offered deals—to loyal customers, to authors, to delivery companies, to resellers agreeing to run ads on their own websites. He drew more and more partners to Amazon. Bezos broke out of the narrow world of books and DVDs and into . . . everything. This kind of experimentation and aggression is what the military calls the OODA loop: "observe, orient, decide, and act." By acting quickly and decisively, you force the enemy—in this case, other retailers—to respond to your last maneuver as you're entering the next one. In Amazon's case, this was done with a ruthless focus on the consumer.

It also helped that, for the better part of Amazon's first fifteen years in existence, traditional retail CEOs were apt to remind people that e-commerce only accounted for 1, 2, 3, 4, 5, 6 . . . percent of retail. There was never a concerted effort to respond to the threat until Amazon had enormous fangs and unlimited capital—it was too late.

Fast-forward to 2016—U.S. retail grew 4 percent, and Amazon Prime grew 40 percent plus.[32,33] The internet is the fastest-growing channel in the largest economy in the world, and Amazon is capturing the majority of that growth.[34] In the all-important holiday season (November and December 2016), Amazon captured 38 percent of online sales. The next nine largest online players captured 20 percent combined.[35] In 2016, Amazon was considered America's most reputable firm.[36]

Zero Sum

With retail growth essentially flat across the American economy, Amazon's growth must be coming from somewhere. Who's losing? Everyone. The graph below, describing ten-year stock appreciation of major U.S. retailers (2006–2016), says it all:

2006-2016 STOCK PRICE GROWTH

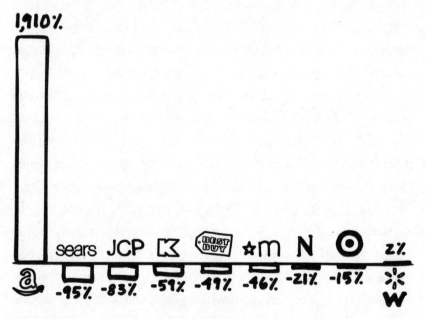

Choudhury, Mawdud. "Brick & Mortar U.S. Retailer Market Value—2006 Vs Present Day." ExecTech.

STOCK PRICE CHANGE ON 1/5/2017

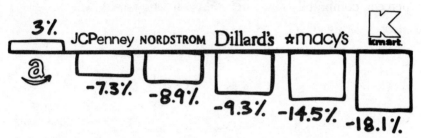

Yahoo! Finance. https://finance.yahoo.com/

Too many stores, flat wages, changing tastes, and Amazon have created the perfect storm for retail. Today, most retailers are getting shelled. Most, but not all.

Amazon has become the Prince of Darkness for retail, occupying a unique position—inversely correlated to the rest of the sector.

Traditionally, stocks in the same sector trade sympathetically—in lockstep with one another. No more. The equity markets now believe that what's good for Amazon is bad for retail, and vice versa. It's a situation almost unique in business history. And it has become a self-fulfilling prophecy, as Amazon's cost of capital declines while every other retailer's increases. It doesn't matter what the reality is—Amazon will win, as it's playing poker with ten times the chips. Amazon can muscle everyone else out of the game.

The real hand-wringing is going to begin when people start asking if what's good for Amazon is bad for society. It's interesting to note that even while some scientists and tech tycoons (Stephen Hawking, Elon Musk) publicly worry about the dangers of artificial intelligence, and others (Pierre Omidyar, Reid Hoffman) have funded research on the subject, Jeff Bezos is implementing robotics

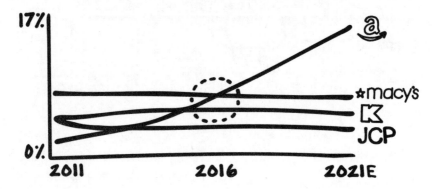

U.S. MARKET SHARES
APPAREL & ACCESSORIES

Peterson, Hayley. "Amazon Is About to Become the Biggest Clothing Retailer in the US." *Business Insider.*

as fast as he can at Amazon. The company increased the number of robots in its warehouses 50 percent in 2016.[37]

With the announcement of Amazon Go, a cashier-less convenience store, the firm entered the brick-and-mortar business. But with a twist: customers at the first Amazon Go groceries-and-goods stores can simply buy items by walking out of the store. Sensors scan your bags, and your app, as you walk out. There's no checkout.

Other retailers, once again rocked back on their heels, are now scrambling to eliminate their own checkout processes. Whom does this latest Amazon maneuver put at risk? The 3.4 million Americans (2.6 percent of the U.S. workforce) employed as cashiers.[38] That's a lot of workers—close to the number of primary and secondary school teachers in the United States.[39]

As retailers are coping with the zig of Amazon Go, hardware

makers, and soon brands, are trying to cope with the zag of Amazon Echo.

Echo is the speaker-like cylinder, and Alexa is its artificial intelligence, named after the library of Alexandria.[40] Alexa is designed to operate like a personal communicator, enabling the user to call up music, search the web, and get answers to questions. Most of all, it takes gathering to the next level by ordering through powerful speech recognition software. Say, "Alexa, add Sensodyne to shopping cart" or (such a pain) push a Trojan Condoms Dash button[41]—and in an hour or less, it's go time. And Alexa gets smarter every time you use it.

That's what the customer gets. For Amazon, the reward is greater: Amazon's customers trust it so much that they're allowing the company to listen in on their conversations and harvest their consumption data. This will give Amazon deeper penetration into the private lives and desires of consumers than any other company.

In the short term, Go and Echo suggest that the company is headed toward zero-click ordering across its operations. Leveraging big data and unrivaled knowledge of consumer purchasing patterns, Amazon will soon meet your need for stuff, without the friction of deciding or ordering. I call this concept Prime Squared. You may need to calibrate every once in a while—less stuff when you go on vacation, more when you're having people over, less Lindt Chocolate when you fall out of love with it—but everything else will operate on retail's equivalent of fly-by-wire. Your order will arrive with an empty box; you'll put the stuff you don't want in the return box, and Amazon will record your preferences. Next time, the return box will get smaller. Amazon made a move in the direction of zero-click ordering when it launched its Wardrobe service in June 2017, allow-ing customers to choose clothes and accessories to try on at home before deciding which to keep. Customers have seven days to make a decision and are only charged after they've made their selection.[42]

Now, compare that to stopping at the shopping center on the way home from work, searching for a parking space, waiting in line only to find that they don't have the kind of lightbulb you were looking for, waiting in yet another line for checkout for your other stuff, and then dealing with traffic on the way home. How will the mall or the box store, much less the mom-and-pop shop, compete? We are witnessing the great reckoning in retail. Just as we witnessed the percentage of our populace working in agriculture decline from 50 percent to 4 percent in a century, we'll see a similar drop over the next thirty years in retail.[43]

Amazon's unwavering focus on making consumer purchases increasingly frictionless, its facility with investor relations, and its decision to invest in B2B (platform services for competitors) place it in the pole position for the race to a trillion. What will cement Am-azon's ownership of the retail world is its commitment, with every move it makes, to gather mountains of data on every consumer in the world. Amazon already knows a great deal about you and me. Pretty soon it will know more about our shopping preferences than we know ourselves. And we're cool with that, as we will have volun-tarily handed over all that information.

Storytelling → Cheap Capital

Amazon has had more access to cheaper capital for a longer period than any firm in modern times.

Most successful VC-backed tech companies in the nineties raised less than $50 million before showing a return to investors. By comparison, Amazon raised $2.1 *billion* in investors' money before the company (sort of) broke even.[44] As the company has shown, Amazon can launch a phone, invest tens, maybe hundreds, of millions of dollars on development and marketing, have it fail within the first thirty days, and then treat the whole disaster as a speed bump.

Now *that* is patient capital. If any other Fortune 500 company— be it HP, Unilever, or Microsoft—launched a phone that proved DOA, their stock would be off 20 percent plus, as Amazon's stock was in 2014.[45] But as shareholders screamed, the CEOs of those other companies would blink and order a company-wide retreat and pull in its horns. Not Amazon. Why? Because if you have enough chips and can play until sunrise, you'll eventually get blackjack.

This cuts to Amazon's core competence: *storytelling.*

Through storytelling, outlining a huge vision, Amazon has reshaped the relationship between company and shareholder. The story is told via media outlets, especially those covering business and tech. Many of them have decided tech CEOs are the new celebrities, and they give Amazon the spotlight, center stage, and star billing anytime. Until now, the contract companies have with shareholders is: give us a few years and tens of millions of dollars . . . and then we'll begin returning capital to you in the form of profits. Amazon has exploded this tradition, replacing profits with *vision and growth*, via storytelling. The story is compelling and simple—the power couple of messaging.

The Story: Earth's Biggest Store.

The Strategy: Huge investments in consumer benefits that stand the test of time—lower cost, greater selection, and faster delivery.

Thanks to a rate of growth that reflects a steady march toward this vision, the market bids Amazon stock higher and provides the firm with exceptionally cheap capital. Most retailers trade at a multiple of profits times eight.[46] By comparison, Amazon trades at a multiple of forty.[47]

In addition, Amazon has trained the Street to hold them to a different standard—to expect higher growth but lower profits. That enables the company to take the (substantial) incremental gross margin dollars it earns each year and plow more capital back into the business—and avoid that whole tax thing. And that in turn funds the digging of deeper and deeper moats around the business.

Profits are to investors what heroin is to an addict. Investors love profits, I mean *really* love them. Yes, invest, grow, and innovate, but don't dare get in the way of me getting jacked up on skag (profits).

Amazon's revolutionary timeline of capital allocation is what has been preached for generations in business school—total disregard for the short-term needs of investors in pursuit of long-term goals. A company that does this is as rare as a young adult who skips prom to study.

Normal business thinking: If we can borrow money at historically low rates, buy back stock, and see the value of management's options increase, why invest in growth and the jobs that come with it? That's risky.

Amazon business thinking: If we can borrow money at historically low rates, why don't we invest that money in extraordinarily expensive control delivery systems? That way we secure an impregnable position in retail and asphyxiate our competitors. Then we can get *really big, fast.*

Walmart wants to impress its parents, and is earnestly investing for the long term. But the markets don't buy this maturity from the Bentonville firm. On Walmart's Q1 2016 earnings call, management informed the Street that the company would be substantially increasing technology capital expenditures to "win the future of retail."[48]

This was the correct, and only, choice for Walmart. However, the strategy meant a reduction in projected earnings. Cue the withdrawals and vomiting. Within twenty minutes of the opening of trading the following day, Walmart's market value shed the equivalent of 2.5 Macy's—$20 billion.[49]

Being an investor in Amazon is growing up in Mitt Romney's house: you're just not going to get access to heroin (profits). Through earnings call after earnings call, Amazon reinforces its vision of growth, downplays profits, and reminds its shareholders that it doesn't *ever* pay dividends. The ointment is a vision of world domination, complete with cool new technologies (drones), content (movies), and *Star Trek* tricorders (Amazon Echo) that have more adoption and buzz than any consumer hardware product since the iPad. It's storytelling, but in a *Harry Potter* way, where the next story is better than the original.

Cheap Capital → 100x Risks

Shrewdly and publicly, Mr. Bezos bifurcates Amazon's risk taking into two types: 1) Those you can't walk back from ("This is the future of the company"), and 2) Those you can ("This isn't working, we're out of here").[50]

Bezos's view is that it's key to Amazon's investment strategy to take on many Type 2 experiments—including a flying warehouse or systems that protect drones from bow and arrow. They've filed patents for both. Type 2 investments are cheap, because they likely will be killed before they waste too much money, and they pay big dividends in image building as a leading-edge company. Shareholders love these stories; it makes them feel like they're part of an exciting adventure. Plus, every once in a while, they actually pan out—and when they do, Amazon has the fuel (capital) to pour on the spark and start a firestorm that sears the competition. The overlooked lesson here, other than having a shit-ton of capital, is the willingness to perform infanticide on initiatives or products that aren't working, thus freeing up capital (in Amazon's case, human capital) to start new crazy initiatives.

My experience in traditional firms is that anything new is seen as innovative, and the people assigned to it, like any parent, become irrationally passionate about the project and refuse to acknowledge just how stupid and ugly your little project has become. As a result, traditional companies not only have less capital to invest but fewer swings at the plate. Amazon demonstrates real discipline around not ramping investment until they know something is working. For all the hype over the last three years about Amazon's entry into brick-and-mortar retail, the sum of their efforts is around two dozen stores. They haven't found a format they feel they can scale.

Bezos, like any great leader, has the ability to explain a crazy idea in a way that makes it seem less crazy but practical. Wait, that's obvious—how did we not think of that? The really crazy shit isn't stupid, it's "bold." Yeah, a floating warehouse sounds crazy the first time you hear of it. Now, ponder the cost of leasing and running

FLOATING WAREHOUSE PATENT PENDING

a traditional terrestrial warehouse. What are its biggest expenses? Proximity and rent, respectively. Now, think again about a floating warehouse. Not so crazy, right?

Bezos's perpetual message is that it's Amazon's nature to swing for the fences on a regular basis. But the analogy is wrong: in baseball, a grand slam only scores four runs. By comparison, the home runs of Amazon Prime and AWS produced several thousand runs when the Seattle firm connected with the ball. As Bezos wrote in Amazon's first annual letter, in 1997, "Given a 10 percent chance of a hundred times payout, you should take that bet every time."[51]

Needless to say, most CEOs don't think this way. Most won't even take risks that have less than a 50 percent chance of success—no matter how big the potential payoff. This is a big reason why old-economy firms are leaking value to new-economy firms. Today's successful companies may have the assets, cash flow, and brand equity, but they approach risk differently than many tech firms that have seen their death. They live for today and acknowledge that great success only comes with significant, even existential, risk.

There is a survivor bias that plagues old-economy CEOs and their shareholders. My nightmare job is the "invisible until you fuck up" position. These jobs are everywhere: IT, corporate treasurer, auditor, air traffic controller, nuclear power plant operator, county elevator inspector, TSA officer. You'll never be famous, but you have a small, and terrifying, chance of being infamous. CEOs of successful old-economy firms have a similar bias—they are "rich until they fuck up."

CEO pay has become so crazy that on a risk-adjusted basis, you're better off staying out of traffic, logging your six to eight years, and retiring rich. However, if you google "biggest mistakes in business history," the majority of results are risks that firms failed to take, such as Excite and Blockbuster passing on acquiring Google and Netflix, respectively.

History favors the bold. Compensation favors the meek. As a Fortune 500 company CEO, you're better off taking the path often traveled and staying the course. Big companies may have more assets to innovate with, but they rarely take big risks or innovate at the cost of cannibalizing a current business. Neither would they chance alienating suppliers or investors. They play not to lose, and shareholders reward them for it—until those shareholders walk and buy Amazon stock.

Most boards ask management: "How can we build the greatest advantage for the least amount of capital/investment?" Amazon reverses the question: "What can we do that gives us an advantage that's hugely expensive, and that no one else can afford?"

Why? Because Amazon has access to capital with lower return expectations than peers. Reducing shipping times from two days to one day? That will require billions. Amazon will have to build smart warehouses near cities, where real estate and labor are expensive. By any conventional measure, it would be a huge investment for a marginal return.

But for Amazon, it's all kinds of perfect. Why? Because Macy's, Sears, and Walmart can't afford to spend billions getting the delivery times of their relatively small online businesses down from two days to one. Consumers love it, and competitors stand flaccid on the sidelines.

In 2015, Amazon spent $7 billion on shipping fees, a net shipping loss of $5 billion, and overall profits of $2.4 billion.[52] Crazy, no? No. Amazon is going underwater with the world's largest oxygen tank, forcing other retailers to follow it, match its prices, and deal with changed customer delivery expectations. The difference is other retailers have just the air in their lungs and are drowning. Amazon will surface and have the ocean of retail largely to itself.

Making Type 2 investments also desensitizes Amazon's shareholders to failure. All of the Four share this—look at Apple and Google with their not-so-secret autonomous vehicle projects, and Facebook with its regular introduction of new features to further monetize its users, which it then pulls back when the experiments don't pan out. Remember Lighthouse? As Bezos also wrote in that first annual letter: "Failure and invention are inseparable twins. To invent you have to experiment, and if you know in advance that it's going to work, it's not an experiment."[53]

Red, White, and Blue

The Four are all disciplined about getting out in front of their skis, taking big, bold, smart bets, and tolerating failure. This failure gene is at the heart of Amazon's and, more broadly, the U.S. economy's success. I've founded or cofounded nine firms and I'm, generously, 3-4-2 (win-loss-tie). No other society would tolerate, much less reward, me. America is the land of second chances, and even if Jeff Bezos is predictably globalist, the culture at Amazon is distinctly red, white, and blue.

Most uber-wealthy people have one thing in common: failure. They've experienced it, usually in spades, as the path to wealth is fraught with risks, and often those risks end up being . . . well, risky. A society that encourages you to get up after being beaned in the head, dust off your pants, step back into the batter's box, and swing harder the next time is the secret sauce for printing billionaires. The correlation is clear. America has the most lenient bankruptcy laws, attracts risk takers, and, as you might guess, has most of them. Twenty-nine of the fifty wealthiest people on the planet live in the United States, and two-thirds of unicorns (private companies with $1 billion–plus valuations) are headquartered here.[54,55]

Sell the Picks

Just as it's better to own the land under a mine, it's also good business to sell picks to the miners. The California Gold Rush proved that was true 170 years ago. Amazon proves it's still true today. Amazon owns a lucrative mine: the firm divides its revenue between retail sales of consumer products (Amazon itself and Amazon

Marketplace) and "Other," the group that holds ad sales from Amazon Media Group and its cloud services (AWS).[56]

Most e-commerce firms can never get to profitability and, at some point, investors tire of a vision that's "reheated Bezos." The firm gets sold (Gilt, Hautelook, Red Envelope) or shutters (Boo.com, Fab, Style.com). A combination of a winner-take-all ecosystem, accelerating customer acquisition, last-mile costs, and a generally inferior (online) experience, makes pure-play e-commerce untenable.

Amazon doesn't escape this fact. But even if Amazon's core business (pure-play e-commerce) is a difficult one for turning a profit, the immense value Amazon has delivered to consumers has created the most trusted, and reputable, consumer brand on the planet.[57,58] Amazon has dominated e-commerce sales volume, but its business model isn't easily replicated or sustained. These days, it's easy to forget that Amazon did not turn its first profit until Q4 2001, *seven years* after its founding,[59] and has dipped in and out of profitability ever since. In the past few years, Amazon has traded on this brand equity, leveraging it to extend into other businesses, and has expanded into other, simply better (more profitable) businesses. Looking back, Amazon's retail platform just may have been the Trojan Horse that established the relationships and brand later monetized with other businesses.

While year-to-year growth for Amazon's retail business ranged from 13 percent to 20 percent from Q1 to Q3 2015, Amazon Web Services—the retailer's network of servers and data storage technology—has grown 49 percent to 81 percent during that same interval. AWS also grew into a significant portion of Amazon's total operating income, from 38 percent in Q1 2015 to 52 percent in Q3 2015.[60] Analysts predict that AWS could reach $16.2 billion in sales by the end of 2017, making it worth $160 billion—more than the company's retail unit.[61] In other words, while the world still thinks of Amazon as a retailer, it has quietly become a cloud company—the world's biggest.

And Amazon isn't stopping at web hosting. Amazon Media Group alone will likely soon surpass Twitter's 2016 revenue of $2.5 billion,[62] making it one of the largest online media properties.[63] Amazon Prime, the most nonexclusive club in America (44 percent of U.S. households[64]), is offering, for $99/year, free two-day shipping, two-hour shipping on select products (Amazon Now), and music and video streaming, including original content.[65] Ideas for content are given the budget for a pilot, and then viewers are asked to vote online for which series get greenlighted.

Amazon, like any sovereign superpower, pursues a triad strategy: air, land, and sea. Can you, Mr. Retailer, get your stuff to your consumer in an hour? No problem. Amazon can do it for you (for a fee), because it's making the investment you can't afford to make—warehouses run by robots near city centers, thousands of trucks, and dedicated cargo planes. Each day, four Boeing 767 cargo planes carry goods from Tracy, California, via an airport in nearby Stockton that was half the size three years ago, to a 1-million-square-foot warehouse that didn't even exist until last year.[66]

In early 2016, Amazon was given a license by the Federal Maritime Commission to implement ocean freight services as an Ocean Transportation Intermediary. So, Amazon can now ship others' goods. This new service, dubbed Fulfillment by Amazon (FBA), won't do much directly for individual consumers. But it will allow Amazon's Chinese partners to more easily and cost-effectively get their products across the Pacific in containers. Want to bet how long it will take Amazon to dominate the oceanic transport business?[67]

The market to ship stuff (mostly) across the Pacific is a $350 billion business, but a low-margin one. Shippers charge $1,300 to ship a forty-foot container holding up to 10,000 units of product (13 cents per unit, or just under $10 to deliver a flatscreen TV). It's a down-and-dirty business, unless you're Amazon. The biggest component of that cost comes from labor: unloading and loading the ships and the paperwork. Amazon can deploy hardware (robotics) and software to reduce these costs. Combined with the company's fledgling aircraft fleet, this could prove another huge business for Amazon.[68]

Between drones, 757/767s, tractor trailers, trans-Pacific shipping, and retired military generals (no joke) who oversaw the world's most complex logistics operations (try supplying submarines and aircraft carriers that don't surface or dock more than once every six months), Amazon is building the most robust logistics infrastructure in history. If you're like me, this can only leave you in awe: I can't even make sure I have Gatorade in the fridge when I need it.

Stores

The final brick in Amazon's strategy for world domination is its use of shitloads of assets piled up online to conquer the retail landscape offline. That's right—I mean stores, those things that were supposed to perish thanks to e-commerce.

The truth is that the death of physical stores has been vastly overstated. In fact, it's not stores that are dying, but the middle class—and, in turn, the businesses that serve that once-great cohort and its neighborhoods. The largest mall owner in the United States is Simon Property Group. Its shares have been hit hard in 2017 after hitting an all-time high in 2016.[69] However, Simon will likely be fine, as it sold properties in middle- and lower-income neighborhoods to focus on wealthy neighborhoods. Forty-four percent of total U.S. mall value, based on sales, size, and quality among other measures, now resides with the top hundred properties, out of about a thousand malls. Taubman Properties, another owner of high-end malls, reports tenants averaged sales per square foot of $800 in 2015, up 57 percent since 2005. Compare that to CBL & Associates Properties Inc., which operates "B" and "C" malls. Its sales per square foot rose just 13 percent, to $374, during that same period.[70]

So, stores are here to stay—if we are careful what stores we're talking about. But so is e-commerce. Ultimately, the real winners will be those retailers who understand how to integrate both. Amazon aims to be that company.

The next retail age will be coined the "multichannel era"—a time when integration across web, social, and brick and mortar is crucial to success. Everything points to Amazon dominating that era as well. I've said for a while that Amazon will open stores—lots of them. It makes sense for them to acquire either a struggling retailer, like Macy's, or a company with a large footprint and vascular system, like a convenience store franchise. Amazon's greatest expense is shipping, and their highest objective is to reach more and more households in less and less time. This is why it made sense for Amazon to acquire Whole Foods, a 460-store franchise[71] that will give Amazon a physical presence in urban centers, where affluent, fast-to-reach consumers live. Amazon has had a decade of selling groceries online without much success,[72] as customers prefer to buy produce and meat in person. Key to success in the multichannel era is knowing which channel to optimize and how to cater to our hunter-gatherer instincts.

As of this writing, in addition to the Whole Foods acquisition,

Amazon is testing its own grocery stores in Seattle and the San Francisco Bay Area. It now has bookstores in Seattle, Chicago, and New York City (with others planned for San Diego, Portland, and New Jersey). Why does Amazon—bookstore killer—need brick-and-mortar bookstores? To sell the Echo, Kindle, and its other goods. Customers want to see, touch, and feel products, Amazon's chief financial officer Brian Olsavsky admitted.[73] The firm is also testing a dozen pop-up retail stores (with a total of perhaps one hundred planned by the end of 2017) targeted at U.S. malls.[74] This is happening even as venerable retailers Macy's and Sears, including its Kmart chain, and mall giants JCPenney and Kohl's have announced plans to shutter hundreds of stores in 2017.[75,76]

Meanwhile, to get a leg up in the multichannel era, brick-and-mortar behemoth Walmart spent $3.3 billion to buy Amazon competitor Jet.com, in what feels like a corporate midlife crisis and $3.3 billion hair plugs. Walmart was frustrated they weren't making progress in online sales, and their frustration was justified. As Amazon marched on, Walmart's e-commerce sales growth had slowed, even flattened.

Jet.com shows that the difference between a dot-bomb and a unicorn is a huckster vs. a visionary, respectively.[77] How can you tell the difference? One has had an exit/liquidity event. Marc Lore, Jet's founder, is that visionary/huckster. Mr. Lore is Jeff Bezos's brother by another mother. Or, if you're a retail worker, they are the spawn of Ayn Rand and Darwin, raised by Darth Maul. Lore is also a banker who turned to e-commerce and chose a low-consideration category that, even better than books, had replenishment built in: diapers.

In 2005, Lore started diapers.com and launched several other categories for parents under the corporate umbrella Quidsi.[78] When Bezos toured the firm, he must have felt at home, recognizing the warehouses close to urban centers staffed by Kiva Robots standing behind a site run by algorithms. Bezos fell hard and in 2011 paid $545 million for Quidsi.[79] For half a billion dollars Amazon bought momentum in key categories, got some great human capital, and took a competitor off the market. But Lore didn't want to work for Jeff Bezos. He wanted to *be* Jeff Bezos. Twenty-four months later he bolted and, with his new wealth, started Jet.com. This must have felt like a half-a-billion-dollar divorce settlement to your husband, who then moves into the house next door and starts fucking your friends.

The ex is still pissed off. In April 2017 Bezos closed Quidsi and laid off many of its employees. Hey, if you leave me, your brother needs to move out of the basement. Maybe Quidsi should have been shut down. But my bet is this was Jeff saying to Marc, "and fuck you too." We forget most of the world's major organizations are run by humans, middle-aged humans, who have enormous egos that ensure they, on a regular basis, make an emotional/irrational decision.

Jet uses algorithms to encourage you to increase the size of your basket by lowering the price based on cost of shipping and how profitable the bundle is. It has an annual membership fee of $50, similar to wholesale club Costco. This was the first company that had the balls to take on Amazon head-to-head and in its first year raised a quarter of a billion dollars. But there was a glitch: the firm and offering made no sense. Jet.com announced soon after launch that they were scrapping the membership model, as business was so strong without it. This is the PR equivalent of turning chicken shit into chicken salad. At the time of Walmart's acquisition, Jet.com was spending $4 million/week on advertising and needed to get to $20 billion in annual sales—more revenue than Whole Foods or Nordstrom—to break even.[80] As traditional consumer marketing wanes in importance at the hands of digital, and better products emerge that consumers can discover using new tools of diligence, entrepreneurs' ability to spin lemons into lemonade to raise ridiculous amounts of capital, position themselves as "disruptive," and sell to an old-economy firm hysterical over its deepening crow's feet, is the new "marketing."

While Walmart attempts to bolt on an e-commerce operation to its existing physical retail infrastructure, Amazon is building and acquiring stores to complement its robust online retail—and is likely to win as a result. Consumers increasingly prefer a channel-agnostic experience, where digital (specifically your smartphone) serves as the connective tissue between consumer, store, and site. The consumer always wins, and she has a choice: Door 1, a great e-commerce experience; Door 2, a great in-store experience; or Door 3, a great site and store experience connected by her mobile phone. The ability to reserve something on her phone, pay later on mobile or desktop, pick it up in store, and never have to wait in a checkout line is damn near unbeatable. Sephora, Home Depot, and department stores already have this kind of multichannel integration.

The future of retail may currently look more like Sephora than Amazon in its current form. However, Amazon has the assets (capital, technology, trust, unrivaled investment in last-mile fulfillment) to realize the multichannel dreams of consumers, and help other retailers get there (for a price) as well.

Ultimately, then, why should Amazon, the king of online retail, get into multichannel retail?[81] Because e-commerce doesn't work, isn't economically viable, and no pure e-commerce firm will survive long term.

On the front end of the e-commerce channel, the cost of customer acquisition continues to rise as consumers' loyalty to brands erodes. You have to keep reacquiring them. In 2004, 47 percent of affluent consumers could name a favorite retail brand; six years later that number dropped to 28 percent.[82] That makes pure e-commerce play increasingly dangerous. Nobody wants to be at the mercy of Google and disloyal consumers.

Amazon has decided it wants off the merry-go-round of high-price acquisition coupled with zero loyalty. That's why the company, via pricing and exclusive content and products, is asking people either to join Amazon Prime or leave. Prime members represent recurring revenue, loyalty, and annual purchases that are 40 percent greater than non-Prime members.[83] If Prime continues to grow at its current rate, and people continue to cut the cord, within the next

AVERAGE MONTHLY SPEND ON AMAZON
U.S. AVG. 2016

PRIME MEMBER	$193

NON-PRIME	$138

Shi, Audrey. "Amazon Prime Members Now Outnumber Non-Prime Customers." *Fortune.*

eight years more households will have Amazon Prime memberships than cable television.[84]

In addition, the cost to build out a robust multichannel offering—which is rapidly becoming the table stakes for survival in retail—is painfully difficult and expensive. Cue Amazon, whose infrastructure is, effectively, building the cable pipe of stuff into the world's wealthiest households. Seventy percent of U.S. high-income households have Prime.[85] Amazon's storefronts will effectively be warehouses that support Amazon's, and other retailers', last-mile problem.

The cost to get you that little black dress from a warehouse to a truck to a plane to a truck to your house, where you're not home, come back the next day, where you try it on and decide to have a guy in a brown uniform take it back to his truck to a plane to a truck to the warehouse, is (*very*) expensive. Amazon's fulfillment costs have grown 50 percent since Q1 2012.[86] That's not sustainable, unless Amazon can garner membership fees and charge others to use its infrastructure . . . which is exactly where the company is headed.

At the apex of its power, Walmart never had its own planes or drones. Overnight delivery firms FedEx, DHL, and UPS have raised their prices an average of 83 percent over the last decade. And since the advent of tracking thirty years ago, there hasn't been much innovation in the overnight space. In sum, these guys are sticking their chins out, and the biggest stone fist is headed their way. DHL, UPS, and FedEx are worth a combined $120 billion.[87] Much of this value will leak to Amazon over the next decade, as consumers trust Amazon more, and the Seattle firm can boast the largest shipper in the United States and Europe—itself—as its first client.

"Alexa, How Can We Kill Brands?"

Amazon's voice technology, Alexa, may shake the ground below both retail and brands. Many of my colleagues in academia and business believe that brand building will always be a winning strategy. They're mistaken. Of the thirteen firms that have outperformed the S&P five years in a row (yes, there's just thirteen), only one of them is a consumer brand—Under Armour. Note: it will be off next year's list. Creative execs at ad agencies and brand managers at consumer firms may soon "decide to spend more time with their families." The sun has passed midday on the brand era.

Brands are shorthand for a set of associations that consumers use for guidance toward the right product. Consumer packaged

PERCENTAGE OF AFFLUENTS WHO CAN IDENTIFY A "FAVORITE BRAND"
☐ 2007/8 ▨ 2014/15

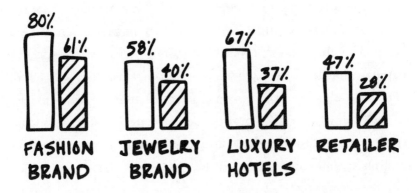

Findings from the 10th Annual Time Inc./YouGov Survey of Affluence and Wealth, April 2015.

goods (CPG) brands like Tide and Coke have spent billions and decades building brand via messaging, packaging, store placement, price, and merchandizing efforts. But when shopping habits migrate online, the design and feel of a product matter much less. There is no visual merchandising, no endcaps with carefully displayed products.

Voice even further circumvents attributes that brands have spent generations and billions to build. With voice, consumers don't know the price or see the packaging and are less likely to include the brand in their request. Fewer and fewer searches contain a brand name.[88] Consumers are willing to price-compare several brands, and Amazon gives them just that opportunity. The death of brand, at the hand of Amazon, and in particular Alexa, can be foreshadowed in search queries.

At L2 we have run tests (by which I mean barking commands at Alexa) to glean insight into Amazon's strategy. Some findings: It's clear that Amazon wants to drive commerce through Alexa, as they are offering a lower price, on many products, if ordered via voice vs. click. In key categories like batteries, Alexa will suggest Amazon Basics, their private label, and play dumb about other choices ("Sorry, that's all I found!") when there are several other brands on amazon.com. Though Amazon carries several brands of batteries, its private label, Amazon Basics, accounts for a third of all battery sales online.

Retailers often leverage their power and custody of the consumer to swap out brands for their own private label. That's nothing new. Only we've never seen any retailer this good at it. Amazon, armed with infinite capital provided by eager investors, is leading a war on brands to starch the margin from brands and deliver it back to the consumer.

Death, for brands, has a name . . . Alexa.

Amazon the Destroyer

I spoke the following morning, after Jeff Bezos, at a recent conference. Similar to the kid who sees dead people in *The Sixth Sense*, Jeff Bezos sees the future of business better than most CEOs. When asked about job destruction and what it would mean for our society, he suggested one more time that we should consider adopting a universal minimum income. Or, he added, a negative income tax where every citizen is granted a cash payment that will be sufficient to stay above the poverty line. People fawned, "What a great man, so concerned about the little guy."

But wait. Ever notice that there are very few pictures of the inside of an Amazon warehouse?

Why is that? Because the inside of an Amazon warehouse is upsetting, even disturbing. Unsafe working conditions? Nope. Abuse of employees as per the *New York Times* article?[89] No. What's disturbing is the absence of abuse, or more specifically, the absence of *people.* The reason Jeff Bezos is advocating a guaranteed income for Americans is he has seen the future of work and, at least in his vision, it doesn't involve jobs for human beings. At least not enough of them to sustain the current workforce. Increasingly, robots will perform many of the functions of human employees, almost as well (and sometimes a lot better), without annoying requests to leave early to pick up their kid from karate.

Amazon doesn't talk publicly about robotics, one of its core competencies, as it realizes it would soon be fodder for late-night hosts and blustery political candidates. In 2012, Amazon quietly acquired Kiva Systems, a sophisticated warehouse robotics firm, for $775 million.[90] In *Star Wars*, Obi-Wan Kenobi feels a dramatic disturbance in the force when the Imperial Army turns the Death Star on Alderaan and destroys the planet. When the acquisition of Kiva closed, every union member should have felt a similar disturbance. Entrepreneurs create jobs, right? No, they don't. Most entrepreneurs, at least in tech, leverage processing power and bandwidth to *destroy* jobs by offering more for less.

Amazon grew its revenues $28 billion in 2016 in a retail environment where growth is essentially flat.[91] If you take the number of people Amazon needs to do one million dollars in revenue vs. the number of people Macy's would need, as Macy's is a decent proxy for retail productivity across the sector (it is, in fact, more productive than most retailers), then it's reasonable to say that Amazon's growth will result in the destruction of 76,000 retail jobs this year. Imagine filling up the largest stadium in the NFL (Cowboy Stadium) with merchandisers, cashiers, sales associates, e-commerce managers, security guards and letting them know that, courtesy of Amazon, their services are no longer needed. Then, be sure to reserve Cowboy Stadium *and* Madison Square Garden next year, as it's only going to get worse (or better, if you are Amazon shareholders).

Amazon isn't unique among the Four in this regard: all do more with less, and all put people out of work.

My first reaction to Bezos's speech was: how refreshing to hear a CEO who didn't quote Ayn Rand. However, as I thought further, I realized that Bezos's words were terrifying. Or just resigned. The guy who has the greatest insight and influence into the future of the world's largest business (consumer retail) has come to the conclusion that there's no way the economy will be able to create, as it has done in the past, enough jobs to replace those being destroyed. Perhaps our society has just given up and doesn't want the burden of trying to figure out how to sustain a middle class.

Ponder that and ask, "Will my kids have a better life than me?"

World Domination

Amazon's path to a trillion likely involves a combination of extension into other parts of the retail value chain and further acquisition. Amazon recently announced it was leasing twenty Boeing 757s, purchasing tractor trailers, and getting into shipping.[92] The doubling of the company's stock in the last eighteen months and the halving of the value of many competing retail stocks (including Macy's and Carrefour) make acquisition an appealing way to add scale and force relationships with brands that have refused to work with them (any luxury brand). The Whole Foods acquisition allows it to establish a foothold in grocery and acquire a few hundred intelligent warehouses currently posing as stores.

Amazon's $434 billion market cap means the Seattle firm could pay (as of April 2016) a 50 percent premium to acquire the outstanding shares of Macy's ($8 billion market cap) and Carrefour ($16 billion) and still only incur an 8 percent dilution to its own shareholders.[93] One can only guess what the U.S. Justice Department would say, but my guess is that it would be happy to make the American economy even more competitive. And the shareholders of Macy's and Carrefour would probably breathe a sigh of relief.

Or, better yet, Amazon could perfect the cashless checkout technology they're working on with Amazon Go, get the media world hot and bothered, and increase the value of the firm by $10 billion. That would be enough to make this, or several other crazy ideas, a reality by throwing cash at it courtesy of the markets, which reward Amazon and punish the rest of retail, as they gaze adoringly at the best storyteller of our age, sans maybe Steven Spielberg—Jeff Bezos.

To be fair, Bezos is delivering on his vision to dominate global retail—and then to own the infrastructure that most consumer businesses will pay a toll to access. European retail growth in 2017 will be 1.6 percent. In 2018, it will be 1.2 percent.[94] Amazon is the top online retailer in Europe, with sales of 21 billion euro in 2015, which beats the next bestsellers, Otto Group and Tesco, by three and five times, respectively.[95]

But the real disruption will occur when Amazon opens stores throughout the rest of the world, as it's planning to in India. People may love Amazon's selections, prices, and the convenience of buying online, but the number-one influencer on consumer decisions is still the store. People love to go into stores and feel things—real, traditional gatherer. This is especially true in grocery, where the instinct first developed. The grocery sector, ripe, certainly, to be disrupted, will see Amazon apply its tech expertise to store logistics, checkout, and delivery, setting new standards in the sector. Whole Foods had been criticized, and its stock price had fallen preacquisition because of its high prices. Amazon will have just the cure for that. Meanwhile the 460 Whole Foods stores become Amazon's supply chain—a delivery hub for Amazon Fresh and a transit hub for its other operations. Whole Foods stores could also become locations for returning online orders of any kind, drastically cutting costs. Amazon wants to be within an hour of as many people as possible, and Whole Foods is a recipe for that.

Imagine if, in the United States, Amazon bought the post office or a gasoline station company. People are used to bombing in and out of these venues to pick up stuff. It's currently building just such "click and collect" stores in Sunnyvale and San Carlos, both in Silicon Valley.[96] That will send a message.

Amazon now offers everything you need, before you need it, delivered in an hour to the 500 million wealthiest households on the planet. Every consumer firm can pay a toll to access an infrastructure less expensive to rent from Amazon than to build itself. Nobody has the scale, trust, cheap capital, or robots to compete. This is all supported by an annual payment that includes all sorts of fun stuff: movies, music, and livestreams of NFL games. My bet is Amazon buys the rights to broadcast March Madness or the Super Bowl to juice their Prime membership . . . as they can.

Race to a Trillion

The circle is now complete. Amazon now has all the pieces in place for zero-click ordering—AI, purchase history, warehouses within twenty miles of 45 percent of the U.S. population, millions of SKUs, voice receptors in the wealthiest American households (Alexa), ownership of the largest cloud/big data service, 460 (soon thousands) brick-and-mortar stores, and the world's most trusted consumer brand.

That is why Amazon will be the first $1 trillion market cap company.

Now, you may ask: What about Apple and Uber? Since 2008, those two companies have created more shareholder value than any other public or private firm. The key to their success was the iPhone and GPS ordering and tracking—and that's very different from Amazon's strategy, right?

Wrong. Their secret sauce was much more mundane: breakthrough stores for Apple and reduced friction for Uber. It's not the GPS tracking illuminating where Javier and his Lincoln MKS are, but your ability to bomb out of the car/store without the friction of paying. That puts both companies on the same playing field as Amazon—and Amazon knows the rules of that game far better than the other two companies.

As Bezos said in his recent letter to stockholders, "At Amazon, we've been engaged in the practical application of machine learning for many years now."[97] How many years? If Amazon tests an AI-like offering anticipating all your retail needs—sending stuff automatically and calibrating based on what you send back or edit via voice ("Alexa, more Rogaine and less sunblock")—the test will register an Amazonian increase in spending per household. The stock will become antigravity matter and triple to a trillion dollars in value. Facebook and Google own media; Apple owns the phone; and Amazon is about to molest the entire retail ecosystem.

Some Big-Ass Losers Here

Retail is a much, much bigger business than media or telco, and Amazon's triumph will mean a lot of losers—not just individual companies, but entire industry sectors.[98,99,100]

INDUSTRY VALUE IN U.S.

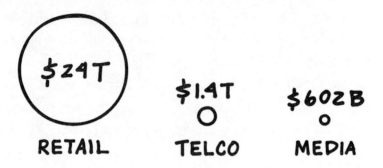

$24T

RETAIL

$1.4T

TELCO

$602B

MEDIA

Farfan, Barbara. "2016 US Retail Industry Overview." The Balance.
"Value of the Entertainment and Media Market in the United States from 2011 to 2020 (in Billion U.S. Dollars)." Statista.
"Telecommunications Business Statistics Analysis, Business and Industry Statistics." Plunkett Research.

Grocery

Obviously, grocery is one of those doomed sectors. It had it coming. This, the largest consumer sector in America ($800 billion[101]), has been where innovation goes to die.[102] Same bad lighting, same depressed workforce, same impossibly frustrating experience in finding my Chobani yogurt as I search from aisle to aisle. Amazon instead offers an online grocery solution with Amazon Fresh and cashier-less grocery shopping with Amazon Go, which debuted in December 2016.[103] In June 2017, Amazon acquired 460 stores in wealthy neighborhoods by way of Whole Foods. While Amazon and Whole Foods account for only 3.5 percent of U.S. grocery spending, the cocktail of high-end grocery and high-tech delivery solutions bodes a significant disruption in the sector. On the day the acquisition was announced, Kroger stock was down 9.24 percent; United Natural Foods, an organic distributor, was down 11 percent; and Target 8 percent.[104] Amazon will eat a lot of lunches.

Restaurants will suffer too, as meal prep at home will be made easier by lightning-fast delivery. And yes, delivery services will take a hit, like Instacart, whose spokesperson said that with the Whole Foods acquisition, Amazon had "declared war on every supermarket and corner store in America."[105]

Walmart

The biggest loser? Easy: Walmart. Walmart's e-commerce growth hurdle reaches beyond Seattle: a workforce that's both underpaid and lacking the skills to close the multichannel circle. Many of their customers are that group you've wondered about, who don't have broadband or a smartphone. The wealthiest man in the twentieth century mastered the art of minimum-wage employees selling you stuff. The wealthiest man of the twenty-first century is mastering the science of zero-wage robots selling you stuff.

The same day that Amazon bought Whole Foods, Walmart bought Bonobos,[106] an online menswear retailer that had acquired brick-and-mortar stores. Bonobos has a strong multichannel model—customers are fitted on site, and clothes are later mailed. Similar to the Jet acquisition, Walmart hopes to derive e-commerce ethos from the smaller retailer, so as to compete against Amazon. Unlikely that Bonobos will make much of a difference, given the scale of the juggernaut.

Walmart is the largest grocery retailer in the United States, and the Whole Foods acquisition is a major escalation in its grocery wars with Amazon.[107] Walmart has ten times the number of grocery stores than Whole Foods, but Amazon's logistics are likely to outsmart it.

Even Google Is Getting Amazoned

Google is, relatively speaking, losing to Amazon. Amazon is Google's largest customer and is better at optimizing search than Google is at optimizing Amazon. Not to say that Google isn't an amazing company, but the good money is on Amazon to beat Google in the race to a trillion. Searches for product are lucrative—they get healthy bids, as there may be a purchase at the end of it, vs. stalking your high-school crush. Amazon's search franchise may rival Google's in value someday, as the people looking to spend start their search at Amazon. But the real victim is traditional retail, whose only growth channel, online, is sunsetting at the hands of Amazon. Each year, Google and brand.coms lose product search volume to Amazon (6 to 12 percent for retailers for 2015 to 2016). Conventional thinking is that consumers are researching on brand sites, then going to Amazon to buy. In reality, 55 percent of product searches start on Amazon (vs. 28 percent on search engines such as Google).[108] This shifts the power, and margin, from Google and retailers to Amazon.

Other Losers: The Unremarkable

I was a remarkably unremarkable kid. I had mediocre grades, but didn't test well either. In high school, I worked as a box boy at The Westward Ho in Westwood, California, and made about $4/hr.

During my freshman year at UCLA I got a job, again as a box boy, this time at Vicente Foods in Brentwood. However, this time, as a member of the United Food & Commercial Workers International Union Local 770, my $13/hr salary paid for my $1,350/year in-state tuition, and then some. Vicente Foods is still there, so it doesn't appear that the 200 percent wage premium that put me through school put Vicente Foods out of business.

In 1984 it was possible to be a remarkably unremarkable kid with a part-time job and pay your way through a tier-1 university. Things have changed a lot, and for kids like my younger self, not for the better. Amazon, good or bad, and the other innovators we worship are making it the best of times for the remarkable, and the worst for the unremarkable.

There will be grocery stores, and box boys, just fewer of them. Like the rest of retail, grocery will bifurcate into "scale" stores with robots giving you 90 percent of a great store for 60 percent of the price, using robotics, cheap capital, software, and voice. These will be stores where the employees are experts and serve the wealthy.

This is our current retail ecosystem. How many of these jobs are likely going to be replaced by more efficient, cost-effective robots? Ask Amazon.

U.S. RETAIL EMPLOYEES
2015

3.4M CASHIERS   **2.8 M** SALESPEOPLE   **1.2M** STOCK CLERKS

"Retail Trade." DATAUSA.

So . . . Is Every Retailer
(and Its Employees) Screwed?

The short answer is no. There is a rebel force of innovative retailers out there who are fighting the empire: Sephora, Home Depot, and Best Buy, to name a few. These firms are zigging as Amazon zags and investing in *people*—beauty associates, blue shirts, geek squads, and gold canvas aprons. They couple this investment in human capital with a deft investment in technology. Consumers no longer go to stores for products, which are easier to get from Amazon. They go to stores for people/experts.

Will their strategy—or Amazon's—eventually emerge victorious? Or will they somehow accommodate each other and carve out a separate peace? The answer will not only decide the fate of companies, but millions of workers and households as well. What's clear is that we need business leaders who envision, and enact, a future with more jobs—not billionaires who want the government to fund, with taxes they avoid, social programs for people to sit on their couches and watch Netflix all day. Jeff, show some real fucking vision.

Chapter 3

Apple

IN DECEMBER 2015, in San Bernardino, California, a twenty-eight-year-old health inspector and his wife attend a work holiday party. They leave their six-month-old daughter with her grandmother. At the party, they don ski masks and fire seventy-five rounds from two variant AR-15 rifles. Fourteen coworkers are killed and twenty-one seriously injured. The assailants die in a shootout with police four hours later.[1] The FBI obtains shooter Syed Rizwan Farook's iPhone 5c and requests—and receives—a federal court order mandating Apple to create and provide software to unlock the phone. Apple defies the order.[2]

Over the next week, I was on Bloomberg TV twice to discuss the issue, and a strange thing happened. I started getting hate mail regarding my view that Apple should comply with the court order. A lot of hate mail.

Wherever you stand in the debate regarding Apple and privacy, the more interesting question is: Would we have endured this hand-wringing if the shooter's phone had been a BlackBerry? No? Why? Because the FBI-inspired court order to unlock the phone would have had a different reception at the door of the Waterloo, Canada, headquarters. My guess is that if the Canadian firm didn't unlock the phone within forty-eight hours, several dozen congressmen and congresswomen would threaten a trade embargo.

Pew surveyed the U.S. public on the issue and found it mostly split. However, there was a huge skew between cohorts. In sum, young Democrats were on Apple's side, and old Republicans, government.[3] That wasn't what you might expect from either side, the former being for expanding the power of big government, and the latter for protecting the prerogatives of big business. But Apple, and the other horsemen, play by a different set of rules.

Put another way, anybody who matters in the consumer world is for Apple. Young Democrats (millennials with college degrees) didn't just inherit the Earth, they conquered it, led by engineering grads from MIT and Harvard dropouts. They are growing their income, spending it irrationally, as young people do, and have a facility with technology that makes them influential and important to business.[4] They sided with Apple, as the firm embodies their own maverick, antiestablishment, progressive ideals—and conveniently ignored the fact that Steve Jobs gave nothing to charity, almost exclusively hired middle-aged white guys, and was an awful person.

It didn't matter, because Apple is cool. Even more, Apple is an *innovator*. And so, when the federal government decides to force Apple to change its behavior, the Apple Macolytes leap to its defense. I'm not one of them.

Double Standard

I've always tried to give the impression that I just don't care what others think. But when coworkers, many of whom are millennials with Ivy League degrees, sent me polite hate mail (which hurts more than just plain hateful, "hope you die," hate mail), it rattled me.

The source of their disappointment was my views on the Apple privacy issue. More specifically, that I wasn't on the correct side of the Apple privacy issue. They felt I wasn't protecting personal privacy. What they failed to see, I believe, is they were not so much on the side of privacy as on the side of Apple. Their, and Apple's, arguments:

- Apple, by creating a new IOS that allowed the FBI to open the phone with brute force, would create a back door that could not be contained and could end up in the wrong hands (SPECTRE?); and
- The government cannot conscript firms into surveillance upon private citizens.

My response to the first claim: If Apple was creating a back door for others to use, it was a pretty unimpressive door. More like a doggy door. Apple estimated that it would take six to ten engineers a month to figure this out.[5] That ain't the Manhattan Project. Apple also maintained this key could end up in the wrong hands and prove hugely dangerous.[6] We aren't talking about the microchip that gave rise to the Terminator, which travels back in time to destroy all humanity. And the FBI even agreed to let the work take place on the Apple campus, ensuring it didn't become an app we can download from www.FBI.gov.[7] Again, these aren't G-men with itchy trigger fingers lurking in doorways and alleys outside the Biograph Theater.

Their second argument, that a commercial firm shouldn't be enlisted in government fights against its will, is a marginally better one. However, does this mean if Ford Motor can construct a car trunk the FBI can't unlock, where it believes there is a kidnap victim suffocating, then the Bureau can't ask Ford to help them get in?

Judges issue search warrants every day. They comply with search-and-seizure laws that prevent indiscriminate searches, and order homes, cars, and computers searched for evidence or information that might prevent or solve a crime. Yet, somehow, we've decided the iPhone is *sacred*. It isn't obliged to follow the same rules as the rest of the business world.

The Sacred and the Profane

Objects are often considered holy or sacred if they are used for spiritual purposes, such as the worship of gods. Steve Jobs became the innovation economy's Jesus—and his shining achievement, the iPhone, became the conduit for his worship, elevated above other material items or technologies.

We thus have, in essence, fetishized the iPhone, and in the process opened the door to a new kind of corporate extremism to emerge. While this extremism doesn't put us in actual physical danger (I don't believe employees of Apple are violent radicals), this kind of secular worship is dangerous. Why? Because when we allow an enterprise to run unchecked and lawless, we've lost respect for the proper standards they, versus other firms, get to play by. The resulting two-tiered system creates a winner-take-all environment that adds further fuel to the flames of inequality. Simply put, Apple in the Steve Jobs era got away with behavior—not least Jobs's own actions regarding backdated stock options awarded to him by Apple[8]—that no other U.S. company CEO would have gotten away with. At some point, the American people, and the U.S. government, decided that Jobs and Apple were no longer constrained by law. Things remained that way until Mr. Jobs's death.

Was it worth it? You decide. In the first decade of the twenty-first century, following Jobs's return to Apple, the company embarked on the greatest run of innovation in business history. In those ten years, Apple introduced one earth-shaking, 100-billion-dollar, category-creating new product or service after another. The iPod, iTunes/Apple Store, iPhone, and iPad . . . there has never been anything like it.

During those years, the consumer electronics industry was a chocolate factory, and Steve Jobs its Willie Wonka. Every winter at the annual Worldwide Developers Conference, Jobs would stand on stage and announce one new product upgrade after another—then start to exit the stage, stop, turn, and say, "Oh, and one more thing . . ." and change the world. Suddenly, what had been a comparatively minor customer convention became an agora. The world's stock markets held their collective breaths. News reporters gathered outside Moscone Center at dawn, previewing the next few hours. And Apple's competitors sat watching newsfeeds, hearts in throats, in terror of what would hit them next.

It's easy to forget now just how stunning Apple's decade was. The iPod's introduction, in late 2001 after the twin shocks of the bursting of the dot-com bubble and 9/11, played the same role as the Beatles' appearance on *Ed Sullivan* just months after the Kennedy assassination: it was a bright light in the darkness that signaled hope and optimism. Then, Jobs used his Hollywood muscle to force an overreaction (that, of course, rewarded Apple) on the audio download piracy, started by Napster, that threatened to destroy the music industry. That set the stage for the masterpiece—the iPhone—that had Apple fanatics all over the world camping out in front of electronics stores. And finally, the sublime iPad. The unsung hero of Apple's success is Napster founder Shawn Fanning, who scared the music industry into the arms of Apple, and who set about partnering with them similar to the way a vampire partners with a blood bag.

Could Apple have maintained this pace into the current decade had Steve Jobs survived his illness? Probably. Because for all of his less than savory traits, he accomplished one important thing: he turned Apple, after the risk-averse years under John Sculley, into a company—arguably the biggest company ever—that made taking risks its *first* option. Unlike every other Fortune 500 CEO, Steve Jobs punished careful thinking, and history recorded the results. Steve Jobs—not Bob Noyce at Intel or David Packard at HP—became the first person to found a company and then make it the most valuable company in the world. Stores, touch screens, and a reheated MP3 player all, at the time, made no sense.

For all the good that Jobs did for Apple, he was also a destructive force inside the company. He bullied employees; his attitudes around philanthropy and inclusiveness were small; his mercurial personality and megalomania kept Apple perpetually in borderline chaos. His death ended the company's historic run of innovation, but it also let Apple, under Tim Cook, focus on predictability, profitability, and scale. You can see the results on the balance sheet: if profits are a sign of success, in fiscal year 2015 Apple was the most successful firm in history, registering $53.4 billion in net profit.[9]

If Apple were anything but a Fortune 500 tech darling, Congress would have implemented tax reforms.[10] But most politicians, like other privileged classes around the world, feel a tiny rush when they pull out their iPhones. It's no contest: Apple—versus, say, Exxon—is likable. C'mon, Think different.

Closer to God

Apple has always found inspiration from others (Latin for stealing ideas). The sector that has inspired Apple's modern-day strategy is the luxury industry. Apple decided to pursue scarcity to achieve outsized, irrational profits that are nearly impossible for new-money, gauche tech hardware brands to imitate. The Cupertino firm controls 14.5 percent of the smartphone market, but captures 79 percent of global smartphone profits (2016).[11]

Steve Jobs instinctively understood this. Attendees at the 1977 Western Computer Conference in San Francisco registered the difference the instant they walked into Brooks Hall: while all other new personal computer companies were offering stripped-out motherboards or ugly metal boxes, Jobs and Woz sat at their table behind the tan injected-plastic Apple II computers that would define the elegant Apple look. The Apple computers were beautiful; they were elegant. Most of all, in a world of hackers and gearheads, Apple's products bespoke *luxury*.

Luxury is not an externality; it's in our genes. It combines our instinctive need to transcend the human condition and feel closer

THE SMARTPHONE GLOBAL MARKETSHARE VS. PROFITS
2016

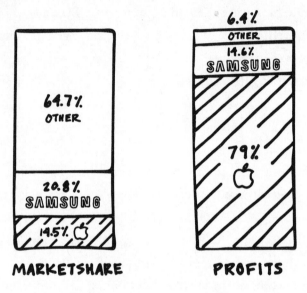

Sumra, Husain. "Apple Captured 79% of Global Smartphone Profits in 2016." MacRumors.

to divine perfection, with our desire to be more attractive to potential mates. For millennia, we've knelt in churches, mosques, and temples, looked around and thought, "There is no way human hands could have created Reims/Hagia Sophia/Pantheon/Karnak. No way mere humans could have created this alchemy of sound, art, and architecture without divine inspiration. Listen to how transcendent the music is. That statue, those frescoes, these marble walls. I'm taken out of the ordinary world. This must be where God lives."

Historically, the masses haven't had access to luxury, so they journeyed to churches and saw chalices encrusted in jewels, gleaming chandeliers, the most beautiful art in the world. They started to associate the combined aesthetic overwhelm from superior artisanship with the presence of God. This is the cornerstone of luxury. Thanks to the Industrial Revolution and the rise of general prosperity, luxury, in the twentieth century, came within reach of hundreds of millions, even billions, of people.

In the eighteenth century, the French aristocracy spent 3 percent of the nation's GDP on beautiful wigs, powders, and dresses. They relied on the opulence of their dress to convey status and inspire respect and submission in their servants. Nike invented neither theater retailing nor endorsements. The Catholic Church has known for centuries the power of an edifice (stores), and built a brand that has survived in the face of wars and astounding scandals. Marie Antoinette's powdered makeup, wigs, and dresses became the rage. Now, Lebron wears Beats. Nothing has changed.

Why? Natural selection—and the desire and envy that arise from it. Powerful people have greater access to housing, warmth, food, and sexual partners. Many who surround themselves with beautiful things claim they are not pursuing a mate, but pure appreciation for the objects. Sort of. The mesh on a Bottega Veneta bag or the slope of the back of a Porsche 911 puts you in the moment. Just. So. Beautiful. You want to possess it, stand in the light of its power, and register how people view you in this softest, most flattering light.

Drive a Porsche, even at fifty-five miles an hour, and you feel more attractive—and more likely to have a random sexual experience. Since men are wired to procreate aggressively, the caveman in us hungers for that Rolex, or Lamborghini—or Apple. And the caveman, thinking with his genitals, will sacrifice a lot (pay an irrational price) for the chance to impress.

Luxury products make no sense on a rational level. We just can't break free of the desire to be closer to divine perfection or to procreate. When luxury works, the act of spending itself is part of the experience. Buying a diamond necklace out of the back of a truck, even if the stones are real, isn't as satisfying as the purchase at Tiffany, from a well-dressed sales assistant who presents the necklace under brilliant lights and speaks in hushed tones. Luxury is the market equivalent of feathers on a bird. It's irrational and sexual, and it easily overwhelms the killjoy, rational signals of the brain—such as "You can't afford this" or "This really makes no sense."

Luxury has also generated enormous wealth. The collision of God and sex atoms ignited energy and value never before seen in business. The list of the four hundred wealthiest people on the planet, minus inherited wealth and finance, includes more people from luxury and retail than technology or any other industry sector. Here's a list of the source of wealth for the ten richest people in Europe (who cares who they are, their companies are infinitely more interesting than they are):

Zara
L'Oréal
H&M
LVMH
Nutella
Aldi
Lidl

Trader Joe's

Luxottica

Crate & Barrel[12]

The Luxury of Time

No technology firm has solved the problem of aging—losing relevance. As a luxury brand, Apple is the first technology company to have a shot at multigenerational success.

Apple did not start as a luxury brand. It was the best house in a shitty neighborhood, tech hardware. A world of cables, geekware, acronyms, and low margins.

In the early days, Apple simply made a more intuitive computer than its competitors. Steve Jobs's notions about elegant packaging only appealed to a minority of customers; it was Steve Wozniak's architecture that drew the rest. Back then, the company appealed largely to consumers' brains. Many early Apple lovers were geeks (which did nothing for its sex appeal). Apple, to its credit, gazed across the tracks at luxury town and thought: Why not? Why can't we be the best house in the best neighborhood?

In the 1980s, the company declined. Machines running Microsoft Windows with Intel chips were faster and cheaper and began to win over the rational organ (the brain). Word and Excel became global standards. You could play most games on the Intel computers, not Apple's. This was when Apple began its move down the torso, from the brain to the heart and genitals—and just in time: the company was destined to sink below 10 percent market share from over 90 percent.[13]

The Apple Macintosh computer, launched in 1984, had attractive icons and a personalized look that appealed to the heart. A computer, it turned out, could be friendly. It talked—at its introduction the computer famously wrote "Hello" on its screen. Artists could express themselves on the Mac, create beauty, and change the world.[14] Then the big breakthrough: desktop publishing. Adobe software was uniquely suited to the Mac's precise, bitmapped display.[15]

Owning an Apple, as embodied in the infamous "1984" commercial, reinforced our belief that Apple users were NOT another brick in the wall.[16] The result was that I, and the employees of my start-ups, struggled through two decades of underpowered and overpriced products just so we could claim we were thinking differently.

But it wasn't sexy. Most people back then didn't go anywhere with their computers. They put them in *computer rooms*. And dragging a potential mate in there to show off some hardware wasn't practical or romantic.

To become a true luxury item, the computer would need to shrink, learn new tricks, be more beautiful, and be in, near, or on your person to signal success to peers in public and private. The transformation began with the iPod, a glossy white block the size of a deck of cards that placed an entire music library in your pocket. Among other mp3 players, all of them awkward gray, navy, and black, the iPod was also a technological miracle—5GB of memory vs. the second largest competitor, Toshiba's 128MB. Apple searched the electronics industry to find a company willing to make a disk drive so tiny, almost jewel-like.

Eventually, Apple would drop "computer" from its corporate name in recognition that the concept of the computer was anchored in the past.[17] The future would be about stuff, from music to phones, *powered* by computers. The customer could carry these branded products around, even wear them. Apple began its march toward luxury.

The 2015 debut of the Apple Watch closed the loop. Its introduction featured on stage a supermodel, Christy Turlington Burns. The cameras panned the audience for gratuitous cameos of famous people. And where did the company buy a seventeen-page spread to celebrate the new arrival? Not in *Computer World*, or even *Time* magazine (as they once had with the Macintosh). No, it was in *Vogue*. And it featured Peter Belanger photos of the rose-gold version, which sells for $12,000. The transformation was complete. Apple had become the best house in the best neighborhood.

Scarcity

A kind of meta-scarcity is key to Apple's success. It may sell millions of iPods, iPhones, iWatches, and Apple Watches, but likely only 1 percent of the world can (rationally) afford them—and that's how Apple wants it.[18] In the first quarter of 2015, the iPhone accounted for only 18.3 percent of the smartphones shipped globally, but 92 percent of the industry's profits.[19] That's luxury marketing. How do you elegantly communicate to friends and strangers that your skills, DNA, and background put you in the 1 percent, no matter where you are? Easy, carry an iPhone.

Plot a heat map of mobile operating systems, and the geography of wealth illuminates. Go to Manhattan. It's all Apple IOS. Head to New Jersey, or out to the Bronx, where average household income plummets, and it's Android. In L.A., if you live in Malibu, Beverly Hills, the Palisades, you have an iPhone. South-Central, Oxnard, and Inland Empire—you own an Android. The iPhone is the clearest signal that you are closer to perfection and have more opportunities to mate.

More writers have written more good articles on Apple than any other company, yet most fail to see it as a luxury brand. I've been advising luxury brands for twenty-five years and believe these firms, from Porsche to Prada, share five key attributes: an iconic founder, artisanship, vertical integration, global reach, and a premium price. Let's dig into each of these more deeply.

1. An Iconic Founder

Nothing builds a self-expressive benefit brand more effectively than the constant personification of the brand in the form of one person, especially the founder. CEOs come and go, but founders are forever. As a poor teen in the 1830s, Louis Vuitton *walked* three hundred miles to Paris, barefoot. He established himself as an expert box maker, and before long was crafting exquisite trunks for the empress of France and wife of Napoleon III, Eugénie de Montijo.[20]

Vuitton was the prototype for the iconic founder. These entrepreneurs have life stories with compelling ups and downs, along with a skill set that is more commonly found in museums than in stores. Art, and the democratization of art (artisanship), fuels and sustains their brands. These founders usually rise from the artisan class. They are blessed/cursed with knowing early what they must do with their lives: make beautiful things. They have no choice.

It's easy to be cynical about bling and the frivolity of the sector. However, drive a Porsche 911, see your cheekbones pop with NARS Orgasm Blush, or find your gaze more intense, your objective more resolute, because you are the guy wearing Brunello Cucinelli.

That's why artisans have created more wealth than any cohort in modern history. "Some people think luxury is the opposite of poverty. It is not. It is the opposite of vulgarity," said Coco Chanel.

To grasp the power of Steve Jobs as the icon for innovation, think of young Elvis. If he had died in his twenties after the Sun Studio sessions and before he left for the army, we never would have seen him waddling across Las Vegas stages in white-bangled bell bottoms. Elvis exited before he hit forty. If he had hung around a few decades longer, he'd be doing oldies acts on retirement cruises, and Graceland would be a mobile home park. Dying removes the icon from the inevitable judgment of everyday existence, including aging, and elevates persona to legend—ideal for a brand. Imagine what the Tiger Woods brand would be worth to Nike if, instead of fading into mediocrity, the once-iconic golf star had been run over by his wife that night she discovered he couldn't keep his putter in his bag. That's arguably one of the few upsides to a public figure passing away—it inoculates them from foolish acts that destroy their reputation and, worse, aging. We know that the Founding Fathers of this country were quietly relieved when George Washington shuffled off this mortal coil—because he was then past the risk of tarnishing his sterling reputation.

It doesn't matter if the iconic founder was a jerk in real life. Apple proves this. The world has created a Jesus-like hero worship of Steve Jobs. In reality it appears that Steve Jobs was not a good person, and a flawed father. He sat in court and denied his own blood, refusing to pay child support to a daughter he knew was biologically his, even though by then he was worth several hundred million dollars. And, as already noted, he also appears to have perjured himself to government investigators regarding the stock option program at Apple.

Yet when Jobs died, in 2011, the world mourned, with thousands posting shrines on the internet, at Apple headquarters, and company stores around the world—and even in front of his old high school. This marked the deification of the iconic founder, moving from stardom to sainthood—a shift made even easier by Jobs's increasingly ascetic look in his final years.

Since then, Apple's brand has burned brighter. There are few better examples of what Pope Francis refers to as an unhealthy "idolatry of money" than our obsession with Steve Jobs. It is conventional wisdom that Steve Jobs put "a dent in the universe." No, he didn't. Steve Jobs, in my view, spat on the universe. People who get up every morning, get their kids dressed, get them to school, and have an irrational passion for their kids' well-being, dent the universe. The world needs more homes with engaged parents, not a better fucking phone.

2. Artisanship

Success in luxury comes from minute attention to detail and expert, almost superhuman, craftsmanship. When it works, it seems as if an alien has arrived from a distant planet to make better sunglasses or silk scarves. Bargain shoppers may wonder why anyone would take so much trouble to design hinges that fold inward, or to knot every tiny thread on the part of a hat that you can't even see. But for people who have discretionary income and aren't worried about survival, the experience of living with a great work of craftsmanship is irreplaceable.

Apple's language of luxury is simplicity, the ultimate sophistication. From the Snow White design style in the eighties (off-white surfaces, horizontal lines to make computers look smaller) to the iPod, "1000 songs in your pocket"—simplicity is an obsession at Apple. Simplicity entails sleek appearance and ease of use—when the interaction with an object sparks delight, brand loyalty increases. The iPod click wheel was at once elegant and playful. The iPhone introduced touchscreens: "You had me at scrolling." Apple chose aluminum for the PowerBook casings, as it was lighter than most materials and allowed for a thinner body and better heat conductivity. And it looks premium and exclusive. As an old iMac ad put it, Apple technology is "Simply amazing, and amazingly simple."[21]

It's how Apple makes products that repeatedly become icons— "objects that appear effortless . . . so simple, coherent and inevitable that there could be no rational alternative."[22] Cognitive psychology shows that attractive objects make us feel good, which in turn makes us more resilient in creative challenges.[23] "Attractive things work better," says Don Norman, vice president of advanced technology at Apple from 1993 to 1998. "When you wash and wax a car, it drives better, doesn't it? Or at least it feels like it does."[24]

3. Vertical Integration

In the early 1980s, The Gap was a chain of pedestrian clothing stores stocked with records, as well as Levi's and other casual clothes, some mixed in with The Gap's own brand. Then, in 1983, the new CEO, Mickey Drexler, remade the company. He softened the lights, bleached the wood, piped in music, expanded the dressing rooms, and decorated the walls with large black-and-white photographs by famous photographers. Each store gave the customer a place to experience the brand Drexler envisioned. He wasn't selling luxury but creating a world around the brand and engaging the consumer face-to-face. He was taking a page from luxury brands and creating a *simulacrum* of that luxury. His strategy stoked revenue and profits, and The Gap began a twenty-year run that was the envy of the retail sector.[25]

Many point to Drexler as "The Merchant Prince." However, his impact on business was greater than that. Drexler recognized that while television could broadcast a brand's message, physical stores could go much further. They gave customers a place to step *into* the brand, to smell it and touch it. The store, Drexler decided, is where he would build brand equity. So, while Gap's key rival, Levi's, continued to create the best TV commercials, Drexler built the best stores.

The result? From 1997 to 2005 The Gap more than tripled in revenue, from $6.5 billion to $16.0 billion, while Levi Strauss & Co. sank from $6.9 billion to $4.1 billion.[26,27,28,29] Brand building moved from the airwaves to the physical world, and Levi's got caught flat-footed. I believe the world would be a better place had LS&Co. registered Apple-like success, as the Haas family (who own LS&Co.) is what you hope all business owners would be: modest, committed to the community, and generous.

Steve Jobs brought Drexler onto Apple's board of directors in 1999, soon after his return to Apple—and two years later Apple launched its first brick-and-mortar store in Tyson's Corner, Virginia.[30] Apple's stores were glitzier than Gap stores. Most experts yawned. Brick and mortar, they said, was the past. The internet was the future. As if Steve Jobs, of all people, didn't understand that.

It's difficult to remember now, but when Apple made that move back then, most people figured the company was wrong; that Apple

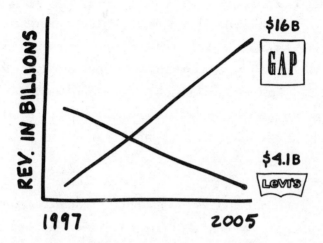

$16B

GAP

$4.1B

Levi's

1997   2005

Gap Inc., Form 10-K for the Period Ending January 31, 1998 (filed March 13, 1998),
    from Gap, Inc. website.
Gap Inc., Form 10-K for the Period Ending January 31, 1998 (filed March 28, 2006),
    from Gap, Inc. website.
"Levi Strauss & Company Corporate Profile and Case Material." Clean Clothes
    Campaign.
Levi Strauss & Co., Form 10-K for the Period Ending November 27, 2005 (filed February
    14, 2006), p. 26, from Levi Strauss & Co. website.

was a company lurching toward irrelevance; and that by opening
fancy stores it was positioning itself for luxury with the equivalent
of a walker. How dumb was that, they thought. Couldn't Apple see
that the tech market now revolved around commodity boxes powered by Microsoft and Intel? That the boom was in e-commerce?

The company's former chief financial officer, Joseph Graziano,
signaled disaster, telling *Business Week* that Jobs was insisting on
"serving caviar in a world that seems content with cheese and crack-
ers."[31]

The stores, of course, changed the tech industry—and advanced
Apple as a luxury company. The iPhone drove Apple's share, but
stores drove the brand and margin. Walk up Fifth Avenue or the
Champs Élysées, and you see Vuitton, Cartier, Hermès, and Apple.

These are captive channels. A $26,000 Cartier Ballon Bleu watch or a $5,000 suede Burberry trench coat would lose their luster on shelves at Macy's. But stores operated by the brands become temples to the brand. Apple's stores sell nearly $5,000 per square foot. Number 2 is a convenience store, which lags by 50 percent.[32] It wasn't the iPhone, but the Apple Store, that defined Apple's success.

4. Global

Rich people are more homogeneous than any cohort on earth. I recently spoke at JPMorgan's Alternative Investment Summit. Its CEO, Jamie Dimon, hosts the bank's three hundred most important (crazy rich) private bank clients and the fifty or so CEOs and founders of the funds into which they invest for their private bank clients. Four hundred masters of the universe, plus people whom the universe has smiled upon (the Lucky Sperm Club). People from nearly every country and culture . . . and yet a sea of sameness. Everyone in the room speaks the same language (literally and figuratively), wears Hermès, Cartier, or Rolex, has kids at Ivy League schools, and vacations in a coastal town of Italy or France or St. Barts. Fill a room with middle-class people from around the world, and you have diversity. They eat different food, wear different clothes, and can't understand each other's languages. It's anthropology on parade. The global elite, by contrast, is a rainbow of the same damn color.

That's why it's easier for luxury brands to permeate geographic boundaries than mass market peers. Mass market retailers, including Walmart and Carrefour, have to hire ethnographers to guide them in local markets. But luxury brands, including Apple, define their own universe. Iconic brand consistency is achieved by key design elements: glass—a glass pane, a cube, or cylinder as an entry, often a clear glass staircase, patented by Jobs; open space, minimal interiors, no inventory in store (products are brought out to purchase). The 492 stores, dropped into exclusive shopping districts in eighteen countries, draw more than 1 million worshippers every day.[33] The Magic Kingdom only drew 20.5 million people total in 2015.[34]

Apple also runs a global supply chain. The components stream in, from Chinese mines, Japanese studios, and American chip fabs, to contractors' immense manufacturing plants and settlements in multiple nations (notably, and notoriously, China), and then on to Apple stores, both brick and mortar and online. Meanwhile, the billions in earnings from the sale of these products follow their own circuitous routes back to a network of tax havens, including Ireland. The result is a gargantuan profit and luxury margins at the scale of a low-cost producer. Apple is one of the most profitable firms in history, but it doesn't need to endure the nuisance of U.S. tax rates.

5. Price Premium

High prices signal quality and exclusivity. Survey your own browsing. Aren't you drawn to, and compelled by, the more expensive item? Even on eBay, don't you search by "highest price" out of curiosity? Negative economic elasticity holds: If Hermès marketed a scarf for $19.95, most existing customers would lose interest. Apple, in this sense, is not Hermès. It can't sell computers or phones for twenty or a hundred times the price of a commodity brand. But it does charge a hefty premium. An iPhone 7 without a subscription subsidy costs $749, a Blu R1 Plus is $159, and the latest from Black-Berry (BlackBerry KeyOne) is $549.[35,36,37]

In this, and in most everything else (except decent HR policies), Steve Jobs learned from Hewlett-Packard, the pioneer in quality tech product pricing. From the first days of Apple Computer, Jobs had publicly stated his admiration of that company and his desire to create Apple in its image. One of HP's attributes that Jobs most admired was its commitment to making the best (that is, most innovative and highest-quality) products—particularly calculators—and then charging the shit out of engineers desperate to buy them. The difference was that HP was largely a professional equipment supplier—hardly a luxury product business—while Apple sold directly to consumers, and thus could take full advantage of all the signals and signifiers of elegance.

Some Apple customers aren't thrilled to learn that their purchases are based on irrational decisions. They think they're smart and sophisticated. So, they rationalize that their brain rode shotgun on the decision. It's just a better phone, they say. The software has an intuitive user interface. And look at all of those cool productivity apps. The laptops work better. The watch encourages me to walk an extra 3,000 steps a day. The higher price is fully justified, they tell themselves.

This may all be true. And people cite similar reasoning when they pay rich premiums for Mercedes or Bentley. Luxury products have to be great. But they also signal status. They improve your procreational brand. This may not be apparent in rich neighborhoods, where it seems that almost everyone carries various Apple gadgets. How cool can you possibly be as the fourteenth person to open a MacBook in Paris's Café de Flore? In these cases, try looking at it the other way. If Apple's the standard, how much does a person's attractiveness to the opposite sex *suffer* when he or she boots up a Dell or pulls out a Moto X to snap a picture?

I'm not saying, by the way, that the sexual bounce coming from a luxury purchase will actually occur. Millions of iPhone owners sleep alone at night. But buying the luxury item triggers an emotion, a boost in serotonin that attends happiness and success. And maybe it does make you more attractive to strangers—certainly a Dell won't. The decision to pay a premium comes from an ancient and primal urge from the lower body—even while the brain yammers on about the rational stuff. (I'll explore this phenomenon further in chapter 7.)

There will be a lot of big losers on the other side of Apple's luxury coin. For example, 2015 was arguably Nike's best year. The firm increased its revenues by $2.8 billion.[38] By comparison, Apple grew its revenues $51 billion.[39] That's an Atlantic Ocean of discretionary dollars people *won't* be spending on other things.

The most likely to tank under the Apple onslaught are the mid-level luxury companies, the ones selling stuff for less than $1,000 (J.Crew, Michael Kors, Swatch, and others). Their customers count their money—and young consumers care more about their phones and coffee than clothes. So, where do limited discretionary dollars go? An old phone with a cracked spiderweb screen limits their options for mating far more than last year's jacket or purse. They might scrimp on the $78 patterned Hedley Hoodie at Abercrombie & Fitch, the $298 quilted-leather shoulder bag at Michael Kors, or the Kate Spade Luna Drive Willow Satchel, which goes for $498.

On the other hand, that $51 billion lost to Apple shouldn't affect platinum brands, such as Porsche or Brunello Cucinelli. Their customers can afford everything, and don't have to choose.

Steve Jobs's decision to transition from a tech to a luxury brand is one of the most consequential—and value-creating—insights in business history. Technology firms can scale, but they are rarely timeless. On the other hand, Chanel will outlive Cisco, and Gucci will witness the meteor that sets Google on a path to extinction. Of the Four Horsemen, Apple has by far the best genetics and, I believe, the greatest chance of seeing the twenty-second century. Keep in mind, Apple is the only firm among the Four Horsemen, at least for now, that has thrived post the original founder and management team.

Tech's Good-Looking Corpse

The research of NYU Stern's Professor of Finance Aswath Damodaran highlights that technology firms experience the traditional company life cycle at an increased speed. They age in dog years, if you will.[40]

The good news is that these tech companies can launch a product, scale the firm, and acquire customers faster than other industries, which face annoyances like real estate, capital requirements or distribution channels that might require years, and a huge labor force, to create. The bad news: the same rocket fuel that sends a tech firm to the Moon also is available to a bevy of younger, smarter, faster competitors coming up fast behind them.

Male lions have a life expectancy of 10–14 years in the wild. However, they live twenty years or more in captivity.[41] Why? Because, in captivity, they aren't constantly challenged by other males. Males in the wild usually die of injuries from fights protecting or challenging the throne. Very few die of old age.

Tech companies are like alpha male lions in the wild. It's good to be king—a higher multiple on earnings, rapid wealth (when it works), and the love and admiration of a society that sees its innovators as rock stars. However, everyone wants to be king. All it takes is strength, speed, violent aggression, and being too stupid to know you will fail, to dethrone the king.

Apple not only transitioned from one of the greatest visionaries to one of the greatest operators—it has been able to extend its life by transitioning to a luxury brand. How? Apple recognized that the CEO after Steve Jobs needed to be an operator who understood how to scale the firm. If Apple's board had wanted a visionary, it would have made Jony Ive CEO.

Vision(less)

I'd argue Apple lacks a vision; however, it still thrives, as making the iPhone bigger and then smaller again is genius in its simplicity (let's take the best bread in the world and slice it a bunch of ways). The firm also has bought more time as it's realized it has the brand, and assets, to make expensive (both capital and time) investments in becoming a luxury brand that other tech firms cannot.

As early as the Macintosh, Apple realized it wanted off the tech train and moved away from the ethos of offering more each year for less money (Moore's Law). Apple's business today is to sell to people goods, services, and emotions—being closer to God and being more attractive. Apple delivers those factors via semiconductor and display technology, powers them with electricity, and wraps them in luxury. It's a potent and intoxicating blend that has created the most profitable company in history. You used to be what you wore, and some now believe you are what you eat. But who you really are has become what you text on.

The Builder King

You would be amazed at how many people still believe, against all evidence, that Steve Jobs actually invented all of Apple's great products. As if he sat at a lab table in the R&D department at Apple headquarters in Cupertino and soldered chips on a tiny motherboard . . . until *boom!* he gave the world the iPod. Actually, that was Steve Wozniak with the Apple 1 a quarter century before.

Steve Jobs was a genius—but his gifts lay elsewhere. And nowhere was that genius more visible than when business experts everywhere were proclaiming the "disintermediation" of tech—the disappearance of the physical distribution and retail channels as they were replaced by the virtualization of e-commerce.

Jobs understood, as none of his peers did, that whereas content, even commodity products, might be sold online, if you wanted to sell electronics hardware as premium-priced luxury items, you had to sell them like other luxury items. That is, in shining temples, under brilliant lights, with ardent young "genius" salespeople at your beck and call. Most of all, you had to sell those items in glass boxes where customers could be seen by others: not just other customers, but passersby, who could peer in and see you among the select. And once you had accomplished that, you could sell almost *anything* in that store—as long as it was elegant, stylishly boxed, and shared the common design tropes with its more expensive peers.

It is why Apple commands margins that no tech company has ever enjoyed, having scaled impossible heights—the premium-priced product, and the low-cost producer. Nothing comes close in other luxury categories. In handbags, Bottega Veneta, a premium-priced handbag, is a high-cost producer. In automobiles, the premium-priced product, Ferrari, is anything but low-cost producer. In hotels, the premium-priced product, Mandarin Oriental, is far from the low-cost producer.

Yet Apple manages to be both . . . and it does so because it emphasized manufacturing and robotics a generation before most tech (especially consumer tech) companies; established a world-class supply chain; and then established a retail presence, backed by a small army of support and IT experts, that has become the envy of every brand and retailer.

Chutes, Ladders, and Moats

Firms try to build higher and higher walls to keep enemies (upstarts and competitors) from invasion. Business theorists call these structures "barriers to entry."

They are nice in theory, but, increasingly, traditional walls are showing cracks, even crumbling—especially in tech. The plummeting price of processing power (Moore's Law again), coupled with an increase in bandwidth and a new generation of leadership that has digital in their DNA, has produced bigger ladders than anyone ever expected. ESPN, J.Crew, and Jeb Bush . . . all unassailable, no? No. Digital ladders (over-the-top video, fast fashion, and @realdonald trump) can vault almost any wall.

So, what's a ridiculously successful firm to do? Malcolm Gladwell, the Jesus of business books, highlights the parable of David and Goliath to make the key point: *don't fight on other people's terms*. In other words, once you've made the jump to light speed as a tech firm, you need to immunize yourself from the same conquering weapons your army levied on the befuddled prey. There are several obvious examples: network effects (everyone is on Facebook because . . . everyone's on Facebook); IP protection (every firm in tech over $10 billion is suing, and being sued by, every other $10 billion tech firm), and developing an industry standard—monopoly—ecosystem (typing this on Word because I have no choice).

However, I'd argue that digging deeper moats is the real key to long-term success.

The iPhone will not be the best phone for long. Too many firms are struggling to catch up. However, Apple has a key asset with a stronger immune system: 492 retail stores in 19 countries.[42] Wait, a marauder could just put up an online store, no? No. HP.com vs. the Apple Regent Street store in London is like bringing a (butter) knife to a gunfight. And even if Samsung decides to allocate the capital, nine women can't have a baby in a month, and the Korean giant would need a decade (at least) to present a similar offering.

Brick and mortar's troubles have been laid at the feet of digital disruption. There is some truth to that. However, digital sales are still only 10–12 percent of retail.[43] It's not stores that are dying, but the middle class, and the stores serving them. Most that are located in, or serving, middle-class households are struggling. By comparison, stores in affluent neighborhoods are holding strong. The middle class used to be 61 percent of Americans. Now they are the minority, representing less than half the population . . . the rest being lower or upper income.[44]

So, Apple, recognizing that ladders will keep getting taller, opted for more analog (time/capital expensive) moats. Google and Samsung are both coming for Apple. But they are more likely to produce a better phone than to replicate the romance, connection, and general awesomeness of Apple's stores. So, every successful firm in the digital age needs to ask: In addition to big, tall walls, where can I build deep moats? That is, old-economy barriers that are expensive and take a long time to dredge (and for competitors to cross). Apple has done this superbly, continually investing in the world's best brand, and in stores. Amazon, also going for moats, is building a hundred-plus expensive and slow-to-get-built warehouses. How old economy! A good bet is Amazon will open thousands before they are done.

Recently Amazon announced leases on twenty 767s and purchased thousands of Amazon-branded tractor-trailers.[45,46] Google has server farms and is launching early twentieth-century aviation technology (blimps) into the atmosphere that will beam broadband down to Earth.[47] Facebook, among the Four Horsemen, has the fewest old-economy moats, making it the most vulnerable to an invading army with big-ass ladders. You can expect that to change, as Facebook announced they, along with Microsoft, are laying cable across the floor of the Atlantic.[48]

The success of single companies like Apple can hollow out entire markets, even regions. The iPhone debuted in 2007, and devastated Motorola and Nokia. Together they have shed 100,000 jobs. Nokia, at its peak, represented 30 percent of Finland's GDP and paid almost a quarter of all of that country's corporate taxes. Russia may have rolled tanks into Finland in 1939, but Apple's 2007 commercial invasion also levied substantial economic damage. Nokia's fall pummeled the entire economy of Finland.[49] The firm's share of the stock market has shrunk from 70 to 13 percent.[50]

What Might Be Next

If you look to the history of Apple and the rest of the Four, each started in a separate business. Apple was a machine, Amazon a store, Google a search engine, and Facebook a social network. In the early days, they didn't appear to compete with each other. In fact, it wasn't until 2009 that Google's CEO at the time, Eric Schmidt, saw the conflict of interest collisions ahead and resigned (or was asked to leave) from Apple's board of directors.

Since then, the four giants have moved inexorably into each other's turf. At least two or three of them now compete in each other's markets, whether it's advertising, music, books, movies, social networks, cell phones—or lately, autonomous vehicles. But Apple stands alone as a luxury brand. That difference presents an immense advantage, providing fatter margins and a competitive edge. Luxury insulates the Apple brand, and hoists it above the price wars raging below.

For now, I see modest competition for Apple from the other horsemen. Amazon sells cut-rate tablets. Facebook is no sexier than a phone book. And Google's one venture into wearable computing, Google Glass, was a prophylactic, guaranteeing that the wearer would never have the chance to conceive a child, as nobody would get near them.

Apple likely has deeper moats than any firm in the world, and its status as a luxury brand will aid its longevity. While the other three companies, the alpha lions on the veldt of high-tech competition, still face the prospect of an early demise, only Apple has the potential to cheat death.

Denting the Universe

The cocktail of low-cost product and premium prices has landed Apple with a cash pile greater than the GDP of Denmark, the Russian stock market, and the market cap of Boeing, Airbus, and Nike combined. At some point, does Apple have an obligation to spend its cash? If yes, then how?

My suggestion: Apple should launch the world's largest tuition-free university.

The education market is ripe, and I mean falling-off-the-tree ripe, to be disrupted. A sector's vulnerability is a function of price increases relative to inflation and the underlying increases in productivity and innovation. The reason tech continues to eat more of

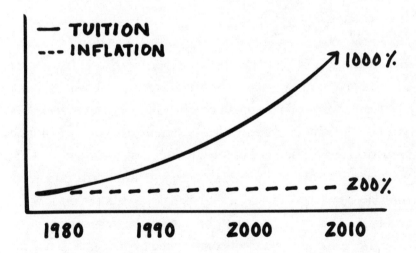

COST OF COLLEGE

— TUITION
--- INFLATION

1000%

200%

1980   1990   2000   2010

"Do you hear that? It might be the growing sounds of pocketbooks snapping shut and the chickens coming home...." AEIdeas, August 2016. http://bit.ly/2nHvdfr.
Irrational Exuberance, Robert Shiller. http://amzn.to/2o98DZE.

the world's GDP is a gestalt that says we need to make a much better product and lower price. Education, on the other hand, has largely remained the same for fifty years and has increased prices faster than cable, and even health care.

I teach 120 kids on Tuesday nights in my Brand Strategy course. That's $720,000, or $60,000 per class, in tuition payments, a lot of it financed with debt. I'm good at what I do, but walking in each night, I remind myself we (NYU) are charging kids $500/minute for me and a projector. This. Is. Fucking. Ridiculous.

A degree from a good school is the ticket to a better life, and this ticket is given almost exclusively to exceptional kids from low- and middle-income U.S. households, and any kid from a wealthy U.S. or foreign household. Eighty-eight percent of kids from U.S. households in the top-income quintile will attend college, and only 8 percent from the lowest. We're leaving the unremarkable and unwealthy—most people—behind in a civilization that is now more Hunger Games than civil.

Apple could change this. With a brand rooted in education, and a cash hoard to purchase Khan Academy's digital framework as well as physical campuses (the future of education will be a mix of off- and online), Apple could break the cartel that masquerades as a social good but is really a caste system. The focus should be creativity—design, humanities, art, journalism, liberal arts. As the world rushes to STEM, the future belongs to the creative class, who can envision form, function, and people as something more— beautiful and inspiring—with technology as the enabler.

A key component would be flipping the business model in education, eliminating tuition, and charging recruiters, as students are broke, and the firms recruiting them are flush. Harvard could foster the same disruption if they take their $37B endowment, cancel tuition, and quintuple the size of their class—they can afford to do this. However, they suffer from the same sickness all of us academics are infected with: the pursuit of prestige over social good. We at NYU brag how it's become near impossible to gain admission to our school. This, in my view, is like a homeless shelter taking pride in how many people it turns away.

Apple has the cash, brand, skills, and market opening to really dent the universe. Or . . . they could just make a better screen for their next phone.

Chapter 4

Facebook

IF SIZE MATTERS (IT DOES), Facebook may be the most successful thing in the history of humankind.

There are 1.4 billion Chinese, 1.3 billion Catholics, and 17 million people who endure Disney World each year.[1,2,3] Facebook, Inc., on the other hand, has a meaningful relationship with 2 billion people.[4] Granted, there are 3.5 billion soccer fans, but that beautiful game has taken more than 150 years to get half the planet engaged.[5] Facebook and its properties will likely pass that milestone before it turns twenty. The company owns three of the five platforms that rocketed to 100 million users the fastest: Facebook, WhatsApp, and Instagram.

You dedicate thirty-five minutes of each of your days to Facebook.[6] Combined with its other properties, Instagram and WhatsApp, that number jumps to fifty minutes. People spend more time on the platform than any behavior outside of family, work, or sleep.[7]

If you believe that Facebook, at $420 billion, is overvalued, then

TIME SPENT ON FACEBOOK, INSTAGRAM, & WHATSAPP PER DAY
DECEMBER 2016

"How Much Time Do People Spend on Social Media?" MediaKix.

imagine if the internet privatized and charged per hour. This Internet, Inc., the company running our digital backbone, then held a public stock offering. What would 20 percent of Internet, Inc.—the typical amount sold in an IPO—be worth? I think $420 billion is low.

Covet

We begin by coveting what we see every day.

—Hannibal Lecter

Facebook is gaining influence faster than any enterprise in history. And that's because what we covet is . . . what's on Facebook. If you

look at the influences that convince a consumer to spend money, Facebook has flooded the awareness stage, the top of the marketing funnel.

What we learn on the social network, and especially on Facebook's subsidiary Instagram, creates ideas and desires. A friend posts an image wearing J.Crew sandals in Mexico, or drinking an Old Fashioned on the rooftop of the Soho House Istanbul, and we want to own/experience these things, too. Facebook gestates intent better than any promotion or advertising channel. Once in pursuit, we go to Google or Amazon to see where to get it. Thus Facebook is higher up the funnel than Google. It suggests the "what," while Google supplies the "how" and Amazon the "when" you will have it.

Historically, in marketing, scale and targeting have been an either/or proposition. The Super Bowl offers scale. It reaches around 110 million people and feeds them nearly identical ads.[8] But the overwhelming majority of those ads are irrelevant to most viewers. You probably don't have restless leg syndrome and are not in the market for a South Korean car. You don't, nor ever will, drink Budweiser. At the other extreme, content presented to a curated group of chief marketing officers, over a dinner hosted by eBay's CMO, is highly relevant to each person at the table. And the dinner for ten costs eBay $25,000+. It's highly targeted, but not scalable.

No other media firm in history has combined Facebook's scale with its ability to target individuals. Each of Facebook's 1.86 billion users has created his or her own page, with years' worth of personal content.[9] If advertisers want to target an individual, Facebook collects data on behavior connected to identities. This is its advantage over Google—and why the social network is taking market share from the search giant. Powered by its mobile app, Facebook is now the world's biggest seller of display advertising—an extraordinary achievement, given Google's brilliant theft of advertising revenues from traditional media just a few years ago.

The irony is that Facebook, by analyzing every bit of data about us, might come closer to understanding us than our friends. Facebook registers a detailed—and highly accurate—portrait from our clicks, words, movements, and friend networks. By comparison, our actual posts, the ones designed for our friends, are mostly self-promotion.

Your Facebook self is an airbrushed image of you and your life, with soft lighting and a layer of Vaseline smeared across the lens. Facebook is a platform for strutting and preening. Users post about peak experiences, moments they want to remember, and be remembered by—their weekend in Paris or great seats at *Hamilton*. Few people post pictures of their divorce papers or how tired they look on a Thursday. Users are curators.

However, the camera operator, Facebook, isn't fooled. It sees the truth—as do its advertisers. This is what makes the company so powerful. The side that faces us, Facebook's users, is the bait to get us to surrender our real selves.

Connecting and Loving

Relationships make us happier. The legendary Grant Study at Harvard Medical School has borne this out. The study—the largest longitudinal study of human beings to date—began tracking 268 Harvard male sophomores between 1938 and 1944. In an effort to determine what factors contribute most strongly to "human flourishing," the study followed these men for seventy-five years, measuring an astonishing range of psychological, anthropological, and physical traits—from personality type to IQ to drinking habits to family relationships to "hanging length of his scrotum."[10] The study found that the depth and meaningfulness of a person's relationships is the strongest indicator of level of happiness.

Seventy-five years and $20 million in research funds, to arrive at a three-word conclusion: "Happiness is love." Love is a function of intimacy and the depth and number of interactions we have with people. At its best, Facebook both taps into our need for these relationships, and helps nourish them. We've all felt it. There's something satisfying in rediscovering someone you knew twenty years ago, and keeping in touch with friends after they move away. When friends post pics of their new baby, we get a delicious hit: dopamine.[11]

As a species, we are weaker and slower than a lot of our competitors. Our developed brain is our competitive differentiation. Empathy is what makes us more human. The explosion in images distributed on social media platforms has led to more empathy, which should make us less likely to gas children, or at least inspire us to hunt down those who do these things. It's common knowledge that countries that trade with one another are less likely to go to war with one another. As deaths from violence continue to decrease (and they are decreasing), I believe we will discover that one of the causes for the decrease is more people feeling closer to . . . more people.[12]

Selflessness and caregiving are key to the survival of the species— and caregivers are rewarded with *life*. The nuance, emotion, and physicality of caregiving keeps us young, as our camera sees we're adding value to humankind. This is Facebook's vital link to our heart, happiness, and health.

A quarter of humanity may populate Facebook feeds with schmaltz and self-delusion. But Facebook also gives users the chance to find love. It turns out that people can send a strong mating signal to their networks just by changing their marital status from "In a Relationship" to "Single." Word of someone's changed status can race through the network, reaching distant nodes that person doesn't know exist.

Facebook analyzes any resulting behavioral changes on the network whenever a customer switches his or her relationship information. As the following graph shows, single people communicate

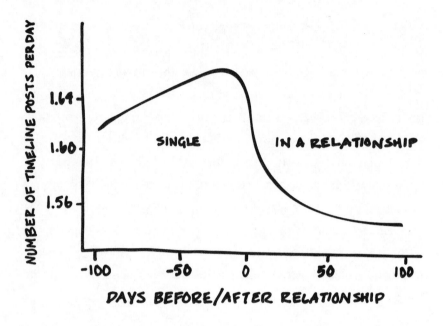

Meyer, Robinson. "When You Fall in Love This Is What Facebook Sees." *The Atlantic.*

more on Facebook. It's part of the preening of courtship. But once they enter a relationship, communication plummets. The Facebook machine tracks this and runs it through a process called "sentiment analysis"—categorizing positive and negative opinions, in words and photos, of each person's level of happiness. And as you might expect, coupling significantly increases happiness (though there appears to be a dip following the initial euphoria).[13]

It's easy to be skeptical about Facebook, especially with all of the self-promotion, fake news, and groupthink spread on the platform. But it's also hard to deny it nurtures relationships, even love. And there is evidence that these connections make us happier.

Watching and Listening

In 2017, one in six people on the planet are on Facebook each day.[14] Users indicate who they are (gender, location, age, education, friends), what they are doing, what they like, and what they are planning to do today and in the near future.

A privacy advocate's nightmare is a marketer's nirvana. The open nature of Facebook, coupled with the younger generation's belief that "to be is to share," has resulted in a data set and targeting tools that make grocery store scanners, focus groups, panels, and surveys look like a cross between smoke signals and semaphore. That data collector behind the two-way mirror, at that focus group that gave you a $75 voucher to Old Navy for participating, is about to lose her job. Simple surveys (and they must be simple, because people today don't have the time for long questionnaires) are near-meaningless in the digital age—when you can measure how people actually behave in their private lives, instead of what they report ("I always use a condom").

This immense learning engine goes well beyond targeting soccer moms on the Nike Page. When you have the Facebook app open on your phone in the United States, Facebook is listening . . . and analyzing. That's right: Anything you do involving Facebook is likely to be gathered and stored.[15] The firm claims it's not using the data to tailor ads, but to better serve up content you may be interested in, or want to share, based on what you are doing (shopping at Target, watching *Game of Thrones*).

What we do know is that Facebook can indeed eavesdrop on ambient noise, picked up on your phone's microphone.[16] That means Facebook can feed this noise into AI-augmented listening software and determine whom you are with, and what you are doing—and even what the people around you are talking about. The targeting isn't any creepier than what happens on the wider web when you have a pixel dropped on your browser and get retargeted ads. That pair of shoes that's following you around the internet? You've been targeted. What's creepy is how good Facebook is getting at it and the number of platforms it can gather and share data across. Double-tap a Vans image on Instagram, and you may find an ad for those same Vans in your Facebook feed the next day. "Creepy" is correlated to relevance.

I don't need to dive too far into the privacy implications here. That discussion is raging on dozens of other channels. But in general, a cold war between privacy and relevance is being waged in our society. No real shots fired yet (like banning Facebook), but both sides (supporting privacy or relevance) don't trust the other, and it could easily escalate. We knowingly feed corporate-run machines a great deal of information about our lives—daily movements, emails, phone calls, the whole package—and then expect firms to make good use of it, but to protect, even ignore, it as well.

Customers, thus far, have indicated that the utility of these platforms is so great that they are willing to endure substantial risks to their data and privacy. Safeguards on networks are insufficient—case in point: Yahoo's data breaches in 2014 and 2016. Data hacks are now deeply, inextricably woven into our lives. I use two-step verification and change my passwords often—I'm told that puts you ahead of people. But I'm still waiting to meet someone who tells me she no longer uses a smartphone or Facebook because of privacy concerns. If you carry a cell phone and are on a social network, you've decided to have your privacy violated, because it's worth it.

The Benjamin Button Economy

Who are the winners in our algorithmically driven economy? Consider a graph. On the y axis is the number of people a company reaches. Facebook and Google, of course, are in the exclusive billion-plus club. But plenty of other companies, from Walmart to Twitter to the TV networks, reach hundreds of millions. On that level, they're superpowers.

But let's put "intelligence" on the x axis. How much does a company learn from its customers? What kind of data do these customers provide? How seamlessly and quickly does it improve the user experience, like auto-populate your destination on Uber, or suggest songs you'll like on Spotify? Over the last five years, only thirteen in the S&P 500 have outperformed the index each year—evidence of our winner-take-all economy.[17] What do most of these firms have in common? They use the peanut-butter-and-chocolate combination

THE NEW ALGORITHM OF VALUE

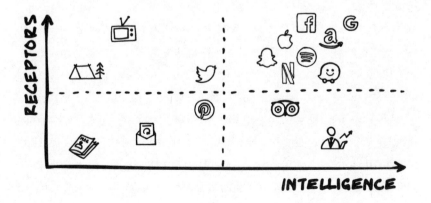

of receptors (users) and intelligence (algorithms that track usage to improve the offering).

This is tantamount to a car that becomes more valuable with mileage. We now have a Benjamin Button class of products that age in reverse. Wearing your Nikes makes them less valuable. But posting to Facebook that you are wearing Nikes makes the network more valuable. This is referred to as "network effects" or "agility." Not only do users make the network more powerful (everyone being on Facebook), but also when you turn on Waze, the service gets better for everyone, as it can geolocate you and calibrate traffic patterns.

Where should you work or invest? Simple: Benjamin Buttons.

Look back at the graph. In the upper right quadrant are the winners, including the three platforms: Amazon, Google, and Facebook. Registering, iterating, and monetizing its audience is the heart of each platform's business. It's what the most valuable man-made things ever created (their algorithms) are designed to do.

Newspapers can reach millions, and many more if you consider how their stories pop up on the three platforms. But they gain almost no intelligence from this contact. Thus, while the three dominant platforms—search, commerce, and social—know me upside down, the *New York Times* has only skeletal details, starting with my address and zip code. It might know I lived in California most of my life. But maybe not. It might try to keep track of my vacation schedule. It sees the stories I read and share, but it's an algorithm targeting a cohort, not a feed-based platform designed specifically for me.

Facebook's algorithm can be used to microtarget distinct populations in specific geographic areas. An advertiser can say, "Give me all the millennial women around Portland looking to buy a car." Using data mined from the social media accounts of millions of

Americans, Cambridge Analytica, a data firm that worked on Brexit and on the Trump campaign, created a "psychographic profile" of voters ahead of the 2016 election. The company used behavioral microtargeting to deliver specific pro-Trump messages that resonated with specific voters for highly personal reasons.[18] With knowledge of 150 likes, their model could predict someone's personality better than their spouse. With 300, it understood you better than yourself.[19]

Like the rest of traditional media, the *Times* let Google handle its search function—until it realized too late its mistake. And so, compared to Facebook, the *Times'* knowledge of me, a fifteen-year subscriber, remains bare bones. TV stations know even less. For the twenty-first century, they're remarkably dumb. And judging by this scheme, dumb companies correlate closely to losers. They were paid to be dumb, as data could have helped advertisers determine which 50 percent of their advertising was wasted and reduce spend.

Some digital companies also lag. Twitter, for example, doesn't know much about its customers. Millions of them have fake names, and as many as 48 million (15 percent) are bots.[20] The result is that while the company can calculate changing moods and appetites in different areas of the planet, it struggles to target individuals. It aces humanity but gets a C in humans. This is the reason Twitter's relevance, similar to Wikipedia or PBS, will always outpace its market value. Good for the planet, bad for Twitter shareholders.

No company is higher or farther to the right on this chart than Facebook. It crushes on both reach and intelligence. This power gives it a huge edge in the digital world. Facebook has access to quinine in a mosquito-infested market—digitally savvy talent. Smart people want to work at a dominant company that they think gets it. Its prospects are bright, opportunities everywhere. There are interesting problems to solve, and ridiculous amounts of money in play. Few firms had the stones, or firepower, to drop $20 billion on a five-year-old company, WhatsApp.

At L2, we track migration patterns between the largest firms, including traditional agencies and the Four. WPP is the world's largest advertising group. Some 2,000 of its former employees have migrated to Facebook or Google. By comparison, only 124 former Facebook or Google peeps left to go work at WPP.

Consider the reverse migrants—124 that went back to WPP. Many of them, it turns out, had only interned at Facebook or Google, and went to WPP when they weren't extended offers in Palo Alto or Mountainside.[21] The ad world today is increasingly run by the leftovers.

This underscores the dominance of the digital giants. It's not

INDIVIDUALS MOVING FROM/TO WPP TO FACEBOOK & GOOGLE

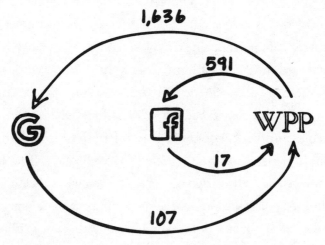

L2 Analysis of LinkedIn Data.

just that their machines are getting smarter, day by day, as they gorge on our data. They attract the best and brightest. Just look at the infamous gauntlet of intelligence tests that America's job seekers are willing to put themselves through for a job at Google. Getting hired at Facebook is no less difficult, just less publicized.

Brains, Brawn, and Blood

Churchill said that WWII was won with British brains, American brawn, and Russian blood. Facebook has all three. If you're wondering which of the three you are, as the customer, it means you're the blood.

Consider Snapchat. Many analysts saw the wildly successful camera app as a potential horseman. A brainchild of Stanford grad students, it stormed out of the gate in 2011, offering a way to send instant photos and videos to friends. The added wrinkle was that videos went poof after a few seconds or hours. It was gaffe insurance, and people felt free to share more intimate content—without worrying about it being seen by a future mate or employer. The ephemeral nature of the content also creates a sense of urgency, resulting in better engagement (cue advertisers salivating). Finally, Snap appeals to teens, a notoriously difficult and influential segment.

Snapchat has added lots of features in the months since its founding. It has even pushed into TV, launching a mobile video channel. In 2017, the company is gaining fast on Twitter, and had 161 million daily users when it filed for an IPO.[22] It IPO'd with a value of $33 billion.[23]

We'll see. Facebook already is positioning itself to crush the young company. Imran Khan, the company's chief strategy officer, claimed: "Snapchat is a camera company. It is not a social company."

I don't know if it's the scorn the Zuck feels after Evan rejected his overtures about acquisition, or a warranted response to a threat. But I believe the first thing Mark Zuckerberg thinks when he opens his eyes in the morning, and the last as he closes them at night, is: "We're going to wipe Snap Inc. off the face of the planet." And he will.

Zuckerberg understands images are Facebook's killer app, much of it residing in the Instagram wing of his social empire. We absorb imagery sixty thousand times faster than words.[24] So, images make a beeline for the heart. And if Snapchat is threatening to hive off a meaningful chunk of that market, or even climb into the lead, that threat must be quashed.

To do this, Facebook is developing a new camera-first interface in Ireland. It's a clone of Snapchat. In a 2016 earnings call, Zuckerberg said, and this may sound oddly similar: "We believe that a camera will be the way that we share."

Facebook has already appropriated (that is, stolen) other Snapchat ideas, including Quick Updates, Stories, selfie filters, and one-hour messages. The trend will only continue—unless the government gets in the way. Facebook is a Burmese python consuming a cow. While the cow goes in, the snake takes its shape. After digesting, it returns to its normal shape, but bigger.

Much of this enormous beast is Instagram. Facebook bought the photo-sharing site in 2012 for $1 billion. It's proving to be one of the greatest acquisitions of all time. In the face of ridicule ("A billion for a company with nineteen people?"), the Zuck was steadfast and pulled the trigger on an asset that's worth fifty-plus times what he paid for it. Whether or not you believe Instagram is the premier platform in its market, it's less of a stretch to acknowledge that it may have been the best acquisition of the last twenty years. (And

Zuckerberg wasn't as lucky two years later—he paid twenty times that for WhatsApp, which had about the same number of employees.)

One way to appreciate the brilliance of this acquisition is to look at Instagram's "Power Index," the number of people a platform reaches times their level of engagement. This social index reveals Instagram as the world's most powerful platform, as it has 400 million users, a third of Facebook's, but garners fifteen times the level of engagement.

Facebook's success with Instagram has a lot to do with its speed in adjusting to the market. Its ability to punch out new features is unrivaled. Some of them work (Messenger, mobile app, customized

GLOBAL REACH VS. ENGAGEMENT BY PLATFORM
Q3 2016

POSTS   **INTERACTIONS**

L2 Analysis of Unmetric Data.
L2 Intelligence Report: Social Platforms 2017. L2, Inc.

news feed), and some fall flat (the snoopy short-lived Beacon, which would share our purchases with our friends, and the failed Buy Button). The birthing, and killing, of new products makes Facebook the most innovative big company on earth.

Less celebrated, but just as important, is Facebook's willingness to quickly back off when it gets pushback from users or the federal government. Facebook knows that its hold on users remains tenuous. Despite the considerable effort those users have put into constructing and maintaining their pages, a sexier competitor could still draw them away by the millions—just as Facebook did to Myspace. So, when its endless monetizing initiatives piss off users—as did Beacon—the company quickly withdraws, waits, then probes somewhere else with some other innovation. Jeff Bezos highlighted in one of his famous investment letters that what kills mature companies is an unhealthy adherence to process. Just ask United Airlines CEO Oscar Munoz, who defended his employees who dragged a passenger off a plane, as they had "followed established procedures for dealing with situations like this."[25]

Much of this innovation comes gratis. Facebook benefits from the ultimate jujitsu move: it will likely become the largest media company on earth, and it gets its content, similar to Google, from its users. In other words, more than a billion customers labor for Facebook without compensation. By comparison, the big entertainment companies must spend billions to create original content. Netflix is shelling out more than $100 million for each season of *The Crown* and will spend $6 billion on content in 2017 (50 percent more than either NBC or CBS).[26] Yet Facebook competes for our attention, and wins it, with pictures of fourteen-month-old Max curled up with his new Vizsla puppy. This is fascinating to a small audience, maybe two hundred or three hundred friends, but that's enough. It's easy for the machine to aggregate, segment, and target. So, to extend the analogy, what would CBS, ESPN, Viacom (MTV), Disney (ABC), Comcast (NBC), Time Warner (HBO), and Netflix (combined) be worth if they had no content costs? Simple—they'd be worth what Facebook is worth.

Duopoly

Google and Facebook are redrawing the media map. Eventually they will control more media spend than any two firms in history—separately, much less combined. Most people would agree that, for the next decade at least, ground zero for growth in media spend will

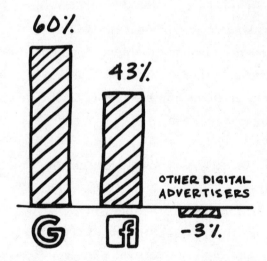

U.S. DIGITAL ADVERTISING GROWTH
2016 YOY

60%

43%

OTHER DIGITAL ADVERTISERS

G   f   -3%

Kafka, Peter. "Google and Facebook are booming. Is the rest of the digital ad business sinking?" *Recode.*

be on mobile. Combined, Facebook and Google control 51 percent of global mobile ad spend, and their share grows every day. In 2016, the two firms accounted for 103 percent of all digital media revenue growth.[27] This means that, sans Facebook and Google, digital media now joins newspapers, radio, and broadcast TV as sectors that are in decline.

Head Fake

As they fight for market dominance, both Facebook and Google can be expected to make bold bets on the future. One especially expensive route leads to virtual reality, and that's where Facebook stole the march on the industry. In 2014, Zuck paid $2 billion for Oculus Rift, the leading VR headset company.[28] Following that acquisition, he raved, "VR will open up new worlds." Spoiler alert: it hasn't.

People were envisioned strapping on headsets to attend virtual work meetings. Surgeons in New York and Tokyo could operate in the same virtual theater. Grandparents would spend virtual time with their far-flung grandkids. In this way, Facebook would get into our heads. It would usher in a new platform—not just for communication, but for spending time together in virtual worlds. The business opportunity was immense.

Following Zuckerberg's lead, venture firms poured hundreds of millions into VR start-ups. Soon, other tech companies, including the Four, were plowing research into the technology. Nobody wanted to sleep through the Next Big Thing.

Virtual reality is the mother of all head fakes. The most powerful force in the universe is regression to the mean. Everyone dies, and gets it wrong along the way. Mark Zuckerberg has been (very) right about a lot of things and was due to make an enormously bad call. And he has. Technology firms do not (yet) have the skills to shape people's decisions on what to wear in public. People care (a lot) about their looks. Most don't want to look like they've never kissed a girl. Remember Google Glass? It got people beaten up. The bottom line is everyone wearing a VR headset looks ridiculous. VR will be to Zuck what Gallipoli was to Churchill, a huge failure that shows he can be (very) wrong, but won't slow his march toward victory. The company is still positioned to dominate the global media market— and reinvent advertising for the twenty-first century.

Insatiable

A devouring beast, Facebook will continue with more of the same. With its global reach, its near-limitless capital, and its ever-smarter data-crunching AI machine, Facebook, in combination with Google, will lay waste to much of the analog and digital media worlds. A decent proxy for what will happen globally to the media business is what has taken place down under, with traditional media being eaten alive by tech media. In sum, old media isn't going away; it will just be a shitty place to work or invest.

A few will hold on. Outfits like the *Economist*, *Vogue*, and the *New York Times* may benefit, at least for a while, because their weaker competitors will die. That, and a sudden recognition that "truth" is a thing again, will give them momentary gains in market share. But the operative word is "momentary."

In the meantime, Facebook will steadily neuter traditional media. The *New York Times*, for example, gets about 15 percent of its online traffic from Facebook.[29] The *Times* agreed to let Facebook post its articles natively on the platform. That means you can read the whole article without leaving Facebook and stepping onto the *Times*' site. The quid pro quo was that the *Times* gets to keep the ad revenues. Sound familiar?

That may sound good, but the reality is that it leaves Facebook in control. That means it can increase or decrease its customers' exposure to the *Times* as it chooses, and swap in and out other media content when Facebook feels like it. This reduces what was once one of the proudest institutions and brands in American media to a commodity supplier. Facebook decides which content is best suited to convey advertising, and who will see it. The *Times* sprayed bullets across their feet letting Google crawl their data. With Facebook Instant Articles, the *Times* and other media firms participating in the program put the gun in their mouth. We have learned nothing. In late 2016, the *Times* pulled from the Instant Articles program, as the revenues were immaterial.[30] So, the *Times* was (again) willing to sell its future, but fortunately the bid wasn't compelling.

Oil

If you drill for oil in certain Saudi Arabian fields, it's pretty simple. You stick a pipe in the ground, and the oil that bubbles to the surface is almost pure enough to pump straight into your car. These can't-miss drilling rigs bring up oil at about $3 per barrel. Even in a depressed market, that same oil sells for about fifty bucks per barrel.

In the heart of America's growing gas belt, in Uniontown, Pennsylvania, a company haggles with a farmer for the mineral rights to his land . . . then drills deep into the earth, hoping to hit a certain type of shale. This company has invested in fancy equipment, with drills that can practically turn corners 10,000 feet underground. It's expensive. And if the company finds the shale, it has to surround it with an industrial production, shattering the rock, pumping in thousands of gallons of briny water, and capturing the natural gas that breaks free. This all costs more than the oil equivalent of $30 per barrel.

Now, would it make sense for Aramco, Saudi Arabia's national oil company, to divert some of its resources to the fracking fields of western Pennsylvania? Of course not, at least for economic reasons. It would give up about $20 per barrel of profit. Why do that?

Facebook faces a similar question. The prime material—the oil—for Facebook is the billions of identities it is following and getting to know in ever-greater detail. The easy money is on the sure things in its people portfolio. By comparison, virtual reality goggles, curing death, laying fiber, self-driving cars, and other business opportunities represent much longer odds. If people make it clear, with their clicks, likes, and postings, that they hate certain things and love others, those people are easy to sell to. Clear as day. Easy as oil in Arabia.

If I go into Facebook and click on an article about Bernie Sanders and "love" one about Chuck Schumer, the machine, expending almost no energy, can throw me in a bucket of liberal die-hards. If it wants to devote a little more computing energy to the process, just to be extra sure, it can see that I have the term *Berkeley* in my bio. So, it delivers me, with great confidence, into the tree-hugger bucket.

The Facebook algorithm then proceeds to send me more liberal pieces, and the company will make money as I click on them. News feed visibility is based on four basic variables—creator, popularity, type of post, and date—plus its own ad algorithm.[31] As I consume that content, whether it's think pieces from the *Guardian*, YouTube clips of Elizabeth Warren expressing outrage at something, or my random friend venting about politics—the algorithm knows what to feed me because it has pegged me as a progressive.

But what about all the people who don't express their politics so clearly? How do you sell political stories to them? Many of them are probably moderates, because most people in America are. And they're a lot harder to figure out. For each one, the Facebook machine would need a much more sophisticated algorithm to analyze their friend network, movements, zip code, the words they use, and the news sites they visit. It's a lot of work, and it's less profitable.

Moreover, after all the work, it's still not a sure thing, because each bucket of moderates to sell to advertisers is based not on direct signals from those individuals, but instead from a host of correlations. Those always come with mistakes. My neighborhood in Greenwich Village is as blue as they come—only 6 percent voted for Trump.[32] Pretty sure that means I'm not just living in a bubble, but a windowless, padded cell. However, as far as windowless, padded cells go, it's pretty nice.

Moderates are hard to engage or predict. Picture a video with some guy in a cardigan sweater discussing, in a balanced tone, the pros and cons of free trade with Mexico. How many clicks would that get? Marketing to moderates is like fracking for gas. You only do it if the easier alternatives aren't available. Thus, we are exposed to less and less calm, reasonable content.

So, Facebook, and the rest of the algorithm-driven media, barely bothers with moderates. Instead, if it figures out you lean Republican, it will feed you more Republican stuff, until you're ready for the heavy hitters, the GOP outrage: Breitbart, talk radio clips. You may even get to Alex Jones. The true believers, whether from left or right, click on the bait. The posts that get the most clicks are confrontational and angry. And those clicks drive up a post's hit rate, which raises its ranking in both Google and Facebook. That in turn draws even more clicks and shares. In the best (worst) cases—we see them daily—the story or clip goes viral and reaches tens or even hundreds of millions of people. And we all step deeper into our bubbles.

This is how these algorithms reinforce polarization in our society. We may think of ourselves as rational creatures, but deep in our brain is the impulse for survival, and it divides the world into us vs. them. Anger and outrage are easily spiked. You can't help yourself but click on that video of Richard Spencer getting punched. Politicians may seem extreme. But they are just responding to the public— and the anger we are working up daily in our news feeds, our march to one extreme.

Clicks vs. Responsibility

Forty-four percent of Americans, and much of the world, turns to Facebook for its news.[33] Yet Facebook doesn't want to be seen as a media company. Neither does Google. The traditional thinking in the market is that they resist this label because of their stock valuations. Why? Because media companies only get a mildly insane valuation, and the Four are addicted to iconospheric valuations— hundreds of billions. That way everyone in their small and select work forces can be not just comfortable, or prosperous, but filthy rich. And that's a retention strategy that is always *en vogue*.

Another reason they don't want to be positioned as media companies is more perverse. Respectable companies in the news business recognize their responsibility to the public and try to come to grips with their role in shaping the worldview of their customers. You know: editorial objectivity, fact-checking, journalistic ethics, civil discourse—all that kind of stuff. That's a lot of work, and it dents profits.

In the case I'm most familiar with, the *New York Times*, I saw that editors not only wanted to get the news right; they tried to achieve a balance in the stories they edited. If there was a bunch of news that seemed to appeal to the left—say, Dreamers being deported or big chunks of Antarctica breaking off and melting—they'd try to get some conservative balance, maybe a David Brooks column attacking Obamacare.

Now people can argue forever about whether the shrinking ranks of responsible media actually achieve balance and get it "right." Still, they try. When the editors are debating which stories to feature, they at least consider their mission to inform. Not everything is clicks and dollars.

But for Facebook, it is. Sure, the company tries to hide this greed behind an enlightened attitude. But basically it's the same MO as the other winners in the tech economy, and certainly the rest of the Four—foster a progressive brand among leadership, embrace multiculturalism, run the whole place on renewable energy—but, meanwhile, pursue a Darwinian, rapacious path to profits and ignore the job destruction taking place at your hands every day.

Don't kid yourself: Facebook's sole mission is to make money. Once the company's success is measured in clicks and dollars, why favor true stories over false ones? Just hire a few "media watchdog" firms to give you cover. As far as the machine sees it, one click = one click. So, entire editorial operations hatch all over the world to optimize production to this Facebook machine. They create crazy fake stories that serve as clickbait for the left and the right.

Pizza Gate—the story about Comet Ping Pong, a pizza parlor in Washington, D.C.—got a lot of momentum around the 2016 election. It claimed that the brother of John Podesta, Hillary Clinton's campaign manager, was running a child prostitution ring in the back rooms, hidden from where the customers eat. Lots of people believed it. One guy drove up from North Carolina with an assault rifle, with vague ideas of freeing the imprisoned and abused children he'd read about. He went into the restaurant and fired a shot, though without hurting anyone (this time), and was arrested.[34]

The shit sandwich here is that having legitimate news next to fake news has only made the Facebook platform more dangerous. When standing in line at Kroger, you may suspect Hilary is not an alien, despite what the *Enquirer* and other supermarket tabloids tell you. However, the presence of the *New York Times* and *WaPo* on Facebook has legitimized fake news.

Platform

How can Facebook exert some form of editorial control? A good place to start is with hate crimes. It's easy to be on the right side of that one. And numerically, the number of people who want to commit hate crimes is not that high. Facebook will raise its hand and say, "No more hate postings!" This way, similar to the rest of the Four, company executives can wrap themselves in a progressive blanket to mask rapacious, conservative, tax-avoiding, and job-destroying behavior that feels more Darwin than (Elizabeth) Warren.

Fake news stories are a far greater threat to our democracy than a few whack jobs wearing white hoods. But fake stories are part of a thriving business. Getting rid of them would force Facebook to accept responsibility as the editor of the world's most (or second most) influential media company. It would have to start making judgments between truth and lies. That would spark outrage and suspicion—the same kind that mainstream media faces. More important, by trashing fake stories, Facebook would also sacrifice billions of clicks and loads of revenue.

Facebook attempts to skirt criticism of its content by claiming it's *not* a media outlet, but a *platform*. This sounds reasonable until you consider that the term *platform* was never meant to absolve companies from taking responsibility for the damage they do. What if McDonald's, after discovering that 80 percent of their beef was fake and making us sick, proclaimed they couldn't be held responsible, as they aren't a fast-food restaurant but a fast-food platform? Would we tolerate that?

A Facebook spokesperson, in the face of the controversy, said, "We cannot become arbiters of truth ourselves."[35] Well, you sure as hell can try. If Facebook is by far the largest social networking site, reaching 67 percent of U.S. adults,[36] and if more us, each day, are getting our news from it, then Facebook has become, de facto, the largest news media firm in the world. The question is, does news media have a greater responsibility to pursue, and police, the truth? Isn't that the point of news media?

As the backlash continued, Facebook introduced tools to help combat fake news. Users can now flag a story as fake, and it will be sent to a fact-checking service. In addition, Facebook is using software to identify potentially fake news.[37] However, with both of those methods, even if false, at most the story is only labeled "disputed." Given the polarization of our political climate and the "backfire effect"—where if you present someone with evidence against their beliefs, they double down on their convictions—a "disputed" label won't persuade a lot of people. It's easier to fool people than to convince them they've been fooled.

We tend to think of social media as neutral—they're just serving us stuff. We are autonomous, thinking individuals and can discern truth from falsehood. We can choose what to believe or not. We can choose how to interact. But research shows that what we click is driven by deeply subconscious processes. Physiologist Benjamin Libet used EEG to show that activity in the brain's motor cortex can be detected 300 milliseconds before a person feels they have decided to move.[38] We click on impulse rather than forethought. We are driven by deep subconscious needs for belonging, approval, and safety. Facebook exploits those needs and gets us to spend more time on the platform (its core success metric is time on site) by giving us plenty of Likes. It sends notifications, interrupting your work or your home life with the urgency that someone has liked your photo. When you share an article that fits your and your friends' political views, you do it expecting Likes. The more passionate the article, the more responses you'll get.

Tristan Harris, former Google design ethicist and expert in how technology hijacks our psychological vulnerabilities, compares social media notifications to slot machines.[39] They both deliver variable rewards: you're curious, will I have two Likes or two hundred? You click the app icon and wait for the wheels to turn—a second, two, three, piquing your anticipation only makes the reward sweeter: you have nineteen Likes. Will it be more in an hour? You'll have to check to find out. And while you're there, here are these fake news stories that bots have been littering the information space with. Feel free to share them with your friends, even if you haven't read them—you know you'll get your tribe's approval by sharing more of what they already believe.

The firm is being careful not to inject humans (gasp!) or any real judgment into the process. It claims that's an effort to preserve impartiality—the same reason it gave when it fired the entire Trends editorial team. To involve humans would supposedly bring on implicit and explicit biases. But AI has biases as well. It's programmed, by humans, to select the most clickable content. Its priorities are clicks, numbers, time on site. AI is incapable of distinguishing fake news, only at best to suspect it, based on origin. Only human fact checkers can ascertain if a story is fake or not, and how high on the scale of credibility.

A digital space needs rules. Facebook already has rules—it famously deleted the iconic image from the Vietnam War of a naked girl running away from her burning village. It also deleted a post by the Norwegian prime minister critical of Facebook's actions. A human editor would have recognized the image as the iconic war photo. The AI did not.

There's a bigger, if unpublicized, reason Facebook as of yet refuses to bring back human editors—it would introduce cost. Why do something the users can do themselves? You get to hide behind freedom of speech, even if you have a crowded theater and someone yells "Fire!" Fear and outrage? All the better. Facebook has good reason not to see itself as a media company. It's too much work and would introduce friction to growth. And that's something the Four don't do.

Utopia/Dystopia

Media platforms where you are the product have empowered, connected, and facilitated greater empathy among billions of people. The shift in value from old-media to new-media firms will result in job destruction and, as with any upheaval, risks.

The greatest threats to modern civilization have come from people and movements who had one thing in common: controlling and perverting the media to their own devices in the absence of a fourth estate that was protected from intimidation and expected to pursue the truth. A disturbing aspect of today's media duopoly, Facebook and Google, is their "Don't call us media, we're a platform" stance. This abdication from social responsibility, enabling authoritarians and hostile actors to deftly use fake news, risks that the next big medium may, again, be cave walls.

Chapter 5
Google

A religion that stressed the magnificence of the universe as revealed by modern science, might be able to draw forth reserves of reverence and awe hardly tapped by conventional faiths. Sooner or later, such a religion will emerge.

—Carl Sagan

MR. SAGAN'S RELIGION IS HERE: it's Google.

Most people, for most of human history, have believed in a higher power. Terrifying weather events led humans to conjure a sentient being orchestrating these phenomena as a response to their behavior. Religion has brought, and still brings, psychological benefits if you're the right candidate. Church, mosque, and temple-goers score higher on optimism and cooperation with one another, which are key paths to prosperity.[1] Believers are more likely to survive than their atheist friends.[2]

However, religion in mature economies is dying. Over the last twenty years in the United States, the number of people who claim no religious affiliation has increased by 25 million. The strongest signal for disbelief is internet usage, accounting for more than a quarter of America's drift from religion.[3] Access to information and education has done a number on belief. People with graduate degrees are less likely to turn to religion than high school graduates.[4,5] You are also less likely to believe in God if you have a high IQ. Only one in six people with an IQ above 140 (uber-smart) report deriving satisfaction from religion.[6]

When Nietzsche proclaimed God is dead, it wasn't a victory cry but a lamentation on the loss of moral compass. As we survive and prosper at greater rates worldwide, what is the glue that holds us together as a human family? What helps us live a better life? How do we learn more, discover more opportunities, find answers to the questions that fascinate and plague us?

Good to Know

Knowledge—we have been fascinated with it since antiquity. Know thyself, admonished the oracle of Delphi. In the Age of Enlightenment, questioning myths became not only okay, but noble—the foundation of liberty, tolerance, progress. Science and philosophy flourished. Religious dogma was challenged with the simple slogan "Dare to know."

More than anything else, we want to know. We want to be sure our spouse still loves us. That our child is safe. Anybody with kids knows the universe collapses to your child, and nothing more, when he or she is ailing. When the kid wakes up with a fever or breaks out in hives, we must know, "Will my universe, my kid, be okay?" The logical part of the brain, the cerebrum, is able to (mostly) calm the reptilian fear brain with facts.

Google answers every question. Our pagan ancestors lived mostly with mysteries. God heard your prayers but didn't answer many of them. If God did speak to you, it meant you were hearing voices, a red flag in any psychological assessment. Most religious people feel watched over, but still don't (always) know what to do. Unlike our ancestors, we are able to find safety in facts. Our questions are answered immediately, our rest assured. How to detect carbon monoxide? Here are five ways. Google even highlights the top answer—here's what you need to know, in big type, in case you're freaking out right now.

Our first instinct is survival. God was meant to provide safety, but only to those who were righteous and denied all their desires. History is replete with believers who begged, fasted, and beat themselves with sticks to implore God for protection and answers. "Is another tribe preparing to attack us?" the oracle at Perperikon would be asked as she poured wine over hot stone. "Who's our greatest enemy?" It was harder to determine North Korea's nuclear head count back then. Now we just type it into the search field.

Prayer

Science has looked for God, or a higher intelligence. Over the last century, there have been numerous well-funded efforts to scan the universe for radio emissions that might register life, for example, the Search for ExtraTerrestrial Intelligence (SETI). Carl Sagan cogently compared this effort to a prayer: lifting your gaze to the heavens, sending up data, and waiting for a response from a more intelligent being. We hope that this superbeing can capture, process, and return an answer.

In the midst of the AIDS crisis, psychiatrist Elisabeth Targ, of the University of California San Francisco, invited psychic healers from as far as 1,500 miles away to pray for ten subjects, each with advanced AIDS. The control group, also ten people, received no prayers from the healers. The results were astonishing and published in the *Western Journal of Medicine*. During the six-month study four subjects died, all from the control group. Dr. Targ did a follow-up study that also showed a statistically significant difference in the levels of CD4+ between test and control groups.

Tragically, Dr. Targ died soon after publishing her research. She was just forty and had been diagnosed with glioblastoma only four months before. She died in furtherance of her research, surrounded by chaos—a cacophony of instructions from shamans, Lakota Sun Dancers, and Russian psychics. After her death, her research didn't hold up to additional scrutiny. Further examination revealed the four patients who died in the original study were the oldest of the twenty subjects. The effectiveness of prayer, the additional scrutiny determined, remains a matter of opinion.[7]

Prayers to Google, however, are answered. It offers knowledge to everyone, despite background or educational level—if you have a smartphone (88 percent of consumers)[8] or an internet connection (40 percent),[9] you can have any question answered. If you want to witness a small part of the staggering diversity of questions asked of Google in real time, go to google.com/about and scroll down to "What the world is searching for now."

Three and a half billion times each day human beings turn their gaze not upward but downward to their screen. We won't be judged for asking the wrong question. Sheer ignorance is welcome—"What is Brexit?" "When is fever dangerous?" Or plain curiosity: "Best tacos in Austin." And we pour out the deepest questions of our heart to our modern-day god: "Why is he not calling me back?" "How do you know if you should get a divorce?"

And answers, mysteriously, appear. Google's algorithms, a work of divine intervention in the eyes of most of us, summon compilations of useful information. The Mountain View search firm answers the questions that plague us, trivial and profound, easing our suffering. Its search results are our benediction: "Go. Take your newfound knowledge and live a better life."

Trust

Apple is considered the most innovative company in the world.[10] Amazon, the most reputable (whatever that means).[11] Facebook is thought of as the best firm to work for.[12] But the trust we place in Google is unrivaled.

One sense in which Google is our modern god is that it knows our deepest secrets. It's clairvoyant, keeping a tally of our thoughts and intentions. With our queries, we confess things to Google that we wouldn't share with our priest, rabbi, mother, best friend, or doctor. Whether it's stalking an old girlfriend, figuring out what caused your rash, or looking up if you have an unhealthy fetish or are just really into feet—we confide in Google at a level and frequency that would scare off any friend, no matter how understanding.

We place immense trust in the mechanism. About one in six

Google queries are questions that have never been asked before.[13] What other institution—professional or clergy—has so much credibility and trust that people bring their previously unanswerable questions to them? What guru is so wise that he inspired so many original questions?

Google bolsters its godlike pose by denoting clearly which search results are organic and which are paid. This boosts confidence in its search, since it seems to be untethered from the marketplace. The result is that Google's scriptures—its search returns—represent for many a stream of unrivaled veracity. Yet Google gets to have it both ways: organic search preserves neutrality, while paid content allows ad revenue. And no one complains.

God is seen as having no agenda when answering queries. He is omnipotent and impartial, loving all his children equally. Google's organic search gives out information that is fair and impartial, with no judgment on who or where you are. Organic search results are based only on relevance to your search terms. Search Engine Optimization can help your site get picked up and appear higher in the list, but SEO is still free and based on relevance.

Consumers trust organic results. We love this impartiality and click on organic results more often than ads. The difference is Google makes money exacting a toll from anybody (Nespresso, Long Beach Nissan, or Keds) that wants to eavesdrop on our hopes, dreams, and worries and present us with ideas on how to address them.

Just as there were personal computers before Apple, online booksellers before Amazon, and social networks before Facebook, there were also search engines before Google. Just Ask Jeeves or Overture. Similarly, just as one or two seemingly minor product features separated the other Four from their packs and turned them into world conquerors—Jobs's design and Wozniak's architecture for the Apple II; the rating and review system for Amazon; photos at Facebook—at Google the defining factors were the elegantly simple homepage and the fact that advertisers weren't allowed to influence search results (organic search).

Neither feature may seem important two decades later, but at the time, they were a revelation. They've gone a long way to creating trust. Google's colorful, uncluttered home page said to even the most neophyte web surfer: "Go for it. Type in anything you want to know. There's no trick involved and no expertise required. We'll take care of *everything*." Meanwhile, when users realized they were getting the best answer, not the one most paid for, it was as if—to continue the biblical analogy—they were seeing the Way, the Truth, and the Light. A bond of trust was created that has survived now for a generation and has made Google the most influential of the Four.

This trust didn't just extend to Google's users, but just as important, to its corporate clients. With Google's auction formula, if advertisers wanted traffic, customers set the prices for each click. If demand drops, so do prices, and you pay just above what someone else was willing to pay, building trust that Google is benign. The result is that corporate customers believe Google's business is run by mathematics, not greed. Once again, the Truth—fair, impartial, constantly calibrated to be equitable.

Compare this trust to the rest of media. Most media outlets, literally and intentionally, do not tell you where the bullshit starts or stops and pretend to have a Chinese wall between editorial and advertising. Some are cleaner than others, but money talks. If you want regular coverage in *Vogue*, then you need to advertise. It's no accident Marissa Mayer got a feature in the magazine, photographed by a top fashion photographer, the same year Yahoo sponsored *Vogue*'s Met Ball.

Yahoo shareholders paid $3 million so Ms. Mayer could appear in *Vogue*.[14] Google, in contrast, keeps that homepage inviolate: it's reserved for search alone, plus the public-service animation of the logo, the Google Doodles. No amount of advertiser money can buy space on the Google homepage. Google anticipated the need for a trust economy in the internet age and helped create it.

In Q3 2016 results, Google had a 42 percent increase in paid clicks. However, the revenue captured (cost per click) declined 11 percent. Analysts mistook this as a negative. Declining prices are typically a reflection of loss of power in the marketplace, as no firm ever willingly drops prices. However, what we missed is that Google was able to grow revenues 23 percent that year and—here's the key part—lower cost to advertisers by 11 percent.[15] Whether you're the *New York Times* or Clear Channel Outdoor, a competitor lowered its prices 11 percent. And word is it's great at what it does and isn't desperate at all. What if BMW was able to improve their cars dramatically each year while lowering prices 11 percent annually? The rest of the auto industry would have trouble keeping up. And, yes, the rest of the media industry, sans Facebook, is having trouble keeping up with Google.

Google found $90 billion in the collection plate in 2016 and commanded $36 billion in cash flow.[16] Several times Congress has debated an incremental tax on firms in sectors that appear to vastly outperform the S&P. Yet nobody has ever suggested extra taxes on

Google. In many religious faiths, to not avert one's eyes from God's face is certain death. The same fate would probably befall the career of any congressman attempting to interfere with Google's progress.

Similar to other horsemen, Google tends to drive prices down, not up. Most consumer firms push in the other direction. They spend a lot of time trying to calculate the maximum price they can charge and capture all excess consumer value. (Booking a same-day flight? Why, you must be a business traveler. Please bend over.) Google works differently, which is why it has grown dramatically year after year after year. And like the other horsemen, it sucks the profits out of its sector. The irony is that Google's victims invited the company in, letting Google crawl their data. Now Google's extraordinary market cap is equal to the next eight biggest media companies combined.[17]

Few people can explain how Google works. Or what Alphabet

MARKET CAPITALIZATION
FEBRUARY 2016

$532B	
	$532B G

$527B						
	$159.6B DISNEP	$141.8B COMCAST	$53B 20th Century Fox	$52.5B Time Warner	$26.8B WPP	$93.3B OTHER

Yahoo! Finance. Accessed in February 2016. https://finance.yahoo.com/

exactly is. Alphabet incorporated in 2015, and Google is one of its subsidiaries in addition to Google Ventures, Google X, and Google Capital.[18] People have an idea about Apple: it builds beautiful objects around computer chips. People understand Amazon: you buy a bunch of stuff at a low price, then people (robots) in a big warehouse pick, pack, and get it to you, fast. Facebook? A network of friends linked to ads. But few people understand what happens inside a holding company that happens to "hold" a gigantic search engine.

Minority Report

The 2002 Tom Cruise film *Minority Report* imagines a world where three mutated humans, "precogs," can see the future and predict criminal acts. The police can then intercede before the crime actually occurs. One of these precogs is better at it than the others—and occasionally sees an alternate future that's hidden. Her better visions are filed away in a "minority report."

Google is a better precog. These are some of the Google queries people have typed in before committing murder, discovered by authorities (unfortunately) post-crime:

"Necksnap Break"

"When someone pisses you off, is it worth it to kill them?"

"Average sentences for manslaughter and murder"

"Fatal digoxin doses"

"Could you kill someone in their sleep and no one would think it was murder?"

The Apple privacy dustup of 2016 will seem trivial as Google's precog powers grow. This will come when a thin layer of AI on top of search queries and a few other data streams, including our movements, are used effectively to predict crime, disease, and stock prices. The information on a smartphone can already put a criminal in prison. But the string of search queries that come barreling out of our lizard brains . . . that's where the really crazy shit can be found. The temptation to create predictive links between intention and action will be irresistible to governments, hackers, and rogue employees.

Look at your recent Google search history: you reveal things to Google that you wouldn't want anyone to know. We believe, naively, that nobody (but the Big Guy) can listen to our thoughts. But let's be clear . . . Google too is listening.

To date, Google has been masterful at keeping this fear in check and not exploiting—as far as we know—the predictive power of its algorithms. Even the company's initial motto, "Don't Be Evil," attempts to reinforce the divine benevolence of this near-supreme being.[19] Moreover, you can be banished: Google has cast out payday lenders, white supremacists, or any firm that charges an interest rate greater than 36 percent. They have been, to recoin a phrase, "cast into outer darkness," the unknown.

But perhaps the greatest sin is to attempt to fool God—that is, game Google's search algorithm. There are 3.5 billion search queries a day,[20] so in essence the search algorithm gets one three-billionth better every time you search.[21] But that's not always the case. In 2011, a *New York Times* investigation revealed that a consultant working for JCPenney had created thousands upon thousands of false links to make it seem as though the JCPenney site was more relevant (that is, had a greater number of other sites linking to it). That false evidence led Google's algorithm to rank the site near the top of its search results, which juiced sales. When the *Times* uncovered this optimization, JCPenney promptly felt the wrath of God. The company was banished to oblivion: second page on Google's search results, the equivalent of being left on the far bank of the Jordan River.[22]

One of God's awesome powers is the knowledge not only of what we do, but also what we *want* to do. We may not have confided in anyone, but as far as many believers are concerned, God knows that as we walk through the mall we lust for a pair of Tory Burch Jolie pumps or Bose QuietComfort headphones. He knows you have a thing for girls with tattoos. Those are temptations God witnesses and registers.

We signal our secret intentions with our queries and provide the Google search engine supernatural power in advertising. Traditional marketing sorted us into *tribes*: Latinos, hicks, retirees, sports fans, soccer moms, and so on. Within those tribes, we are thought to be the same. In 2002, every single rich white suburbanite wore cargo pants, listened to Moby, and drove an Audi. But with Google, our queries—along with the photos, emails, and all the other data we provide—identify us as individuals with distinct problems, goals, and desires. This intelligence gives God a leg up in the advertising business. It can serve up ads that are more relevant, more benevolent—tailored to our personal happiness.

Much of marketing is the art (disguised as science) of how to best change behavior. It wants to make us buy this vs. that, think of this as cool vs. that as lame. Google leaves the hard, expensive stuff to others and just gives the people what they want after they've raised their digital hand and said, "I want something like this." Even better, Google pairs people with companies, via AdWords, before they may even know what they want (fly Delta) when signaling their intent, via search queries about tours of the Acropolis, or the slightest curiosity about "Greek islands."

The Old God

If Google is the god of information in the internet age, the closest thing we had in the old economy, with maybe the exception of the evening news, is the *New York Times*. Its longtime motto—"All the News That's Fit to Print"—says what the paper aspires to. Every day it renders judgment on what's important, on what we should know. Of course, the *Times* has its prejudices—every human institution does. But *Times* journalists pride themselves on keeping these judgments (somewhat) in check. They see themselves upholding progressive Western values—and steer us away from the news that's *not* fit to print, whether porno or propaganda or advertising disguising itself as news.

The editors at the *Times* curate our view of the world we inhabit. When *Times* editors choose the stories for their front page, they set the agenda for TV and radio news, for the whole mainstream view of the world. The stories circulate across the Old World (40 percent of the leaders of nations receive some version of the *New York Times* each morning) and the New (Facebook and Twitter).

Journalism is hard, sometimes dangerous, work that pursues truth vs. just the commercial. The *New York Times* does it better than any media firm in the world. However, increasingly, the paper is not good at extracting value from the skills and risk taking demonstrated in the newsroom.

In fact, Google and Facebook do a better job extracting value from *Times* journalists than the management of the *Times*. I believe if the *Times* had refused to let any of its content on the Facebook or Google platforms, those younger companies would be worth at least 1 percent less. *New York Times* articles give these platforms tremendous credibility, and the *Times* in return gets . . . very little.

NY(low)T

In 2008, the gap between the growing Google and the flagging *New York Times* was smaller than today. Google already was well into its stride, with a market cap that topped $200 billion. But the *Times* was enormously *relevant*.[23] With the first iPhones arriving and tablets three years into the future, platforms and devices needed content—and the *Times* had the best. Without *New York Times* content, Google would have been at a disadvantage to anybody who had it—not least the *Times* itself.

I felt the *Times'* content could, and should, be worth billions in the digital age. Working with two NYU Stern students with finance backgrounds, we evaluated every aspect of the Times Company. Our conclusion: the Times Company was a $5 billion firm trapped in a $3 billion body. I approached Phil Falcone, founder of Harbinger Capital Partners. I'd partnered with Phil before. When I say "partnered," I mean his fund provided the capital for us to take a large ownership stake, obtain board seats, and advocate for change.

Phil was one of twelve children, raised in Minnesota. He had been a hockey star at Harvard before becoming a hedge fund manager. A focused introvert, Phil was one of half a dozen investors with balls of steel who made a huge bet against the credit markets in 2006. This made billions for Phil and his investors. Harbinger's office had bad cherry wood trim, artificial plants, and old floor fans to keep the trading floor cool. It felt like a Regus office suite in Cleveland, minus the charm.

I presented the idea to Phil. It involved both a surrender and a fight. I proposed that the Times Company could sell 10 percent to former Google CEO Eric Schmidt and make him CEO of the paper. This was the surrender. I figured Schmidt could afford to buy 10 percent plus of the firm to give him a vested interest/upside. Eric had kicked himself upstairs to chairman of Google—making way for Larry Page to be CEO.

I believed he was likely more open to a different idea (saving American journalism) than he may have been in the past. The stake would give him a chance to make money, though nothing on the scale of one of the Four. (I'm still convinced that if the *Times* had named Schmidt its CEO, the company's value would be dramatically larger.)

The next step, I continued, was for the company to fight. The *Times* should immediately turn off Google—and henceforth the company should refuse to allow Google, or any other company, to crawl its content. Then, if Google or another internet player wanted to license the content of the *New York Times*, it would have to pay for it—and pay more than anybody else. Google, Bing, Amazon, Twitter, or Facebook could provide their users with unfettered access to our content. But only one of them—the highest bidder.

Then, my plan was to stretch this strategy beyond the *Times*.

I envisioned creating a consortium of newspaper owners—the Sulzbergers of the *Times,* the Grahams of the *Washington Post*, the Newhouses, the Chandlers, Pearson, and Germany's Axel Springer, among others. This group would represent the highest-quality, most differentiated media content in the Western world.

This was our one and only chance to staunch the decline of print journalism and capture (back) billions in shareholder value. It wouldn't have lasted forever. But for an also-ran search engine like Microsoft's Bing, it could have provided a potent weapon against Google. Bing at that point had about 13 percent share of search. Exclusive rights to differentiated content via iconic brands, whether from the *Times*, the *Economist,* or *Der Spiegel,* had to be worth a few points of market share. This means of differentiation was worth billions.

Today, the search industry is worth half a trillion dollars. Some would argue more, as Amazon is technically a search engine with a warehouse attached. That means each point of market share in this industry is worth $5 billion plus. My plan was to form the consortium, lease our content, and start pushing back on tech firms that had built billions in stakeholder value based on our content.

Even as the housing bubble showed signs of strain, and advertising continued to drift online, the newspaper business was robust, and properties were in play. Rupert Murdoch had just bought the *Wall Street Journal* for $5 billion, and the *Times* was trading at a much lower multiple.

In addition, there were other buyers sniffing around. I had heard from two different sources that Michael Bloomberg was also contemplating a bid for the *Times*. Term limits, it seemed, were about to force him from office, and the *Times* was the perfect project for a New York billionaire who had taken financial information, brought it into the digital age, and created tens of billions of shareholder value in the process. (We didn't know at the time that when you are Michael Bloomberg, "term limit" is more of a suggestion than a real limit. Bloomberg went on to strong-arm the city council into a third term.)

Finally, if all else failed, the New York Times Company owned a bunch of stuff we should, and would, sell, including:

- The seventh tallest building in America
- About.com
- 17 percent of the Boston Red Sox (wtf?)

These assets were treated by the financial markets as newspaper assets, meaning their valuation was a multiple of profits assigned to a newspaper company (low). So, the disposition of these assets would be accretive to shareholders. A sum-of-the-parts analysis reflected that in buying a share of Times Company stock, you were getting the paper for almost nothing, based on the value of other assets.

We would also lobby to kill the dividend, a payout for shareholders of around $25 million a year. The company needed the cash to reinvest in innovation. Best as I could tell, the dividend was merely protection money so Arthur Sulzberger and Dan Golden wouldn't be killed at family get-togethers because they were being paid $3–5 million per year to fuck up granddad's company and have lunch with Boutros Boutros-Ghali. The other cousins wanted their sugar, too.

Phil's firm, Harbinger Capital, and Firebrand Partners (the name I gave my firm) teamed up to purchase $600 million in Times

Company stock—about 18 percent of the company. This made us the largest shareholder. We announced we wanted four board seats and would push for shareholders to vote a slate of like-minded reformers to the board. We wanted the company to sell noncore assets and double down on digital. Harbinger was the brawn (capital); Firebrand was the brains (lead the proxy fight, join the board, influence capital allocation decisions and strategy, unlock value, etc.).

Within the company, our plan met with resistance, of course. During our initial meeting with management, and after we laid out our thoughts, Arthur Sulzberger indignantly pronounced, "There isn't a single thing you've presented we haven't thought of!" Despite this, we weren't convinced management didn't need help. Outside the walls of the *Times* building on 41st Street (a tower designed by Renzo Piano I was eager to unload), all hell broke loose. I had underestimated how fascinated the media was with . . . itself. Within twenty-four hours of us announcing our strategy, there were paparazzi outside my class at NYU.

The media also enjoyed beating up on the *Times'* publisher and chairman, Arthur Sulzberger. One Reuters reporter, who was working on a story about the dynamics of the Sulzberger family, called me on my cell one night at eleven. He told me he would be fired the next day unless I gave him something, *anything*, for a story about our battle with the *Times*.

He had assembled, with art supplies, an elaborate family tree—cousins, second cousins—with a level of detail that was downright creepy. It became clear that the world's media couldn't wrap its head around how it felt about the people who owned the media.

Arthur Sulzberger and I took an immediate, almost visceral, disliking to one another. We saw the world differently and approached it from entirely different angles. My whole life has been a quest to gain relevance and fear of never achieving it, whereas Arthur's biggest fear (I believe) was losing it. And to be clear, he was the CEO. He gave Janet Robinson the title just so he wouldn't have to do the shit-work of a CEO—firing people, earnings calls, etc. However, he made the big decisions and collected CEO-level compensation.

The Sulzbergers, like many media families, employ a dual-class shareholder structure to keep them in control. The thinking is that media plays a special role in our society and should not be subject to the short-term thinking of shareholders. Most use this (Google, Facebook, Cablevision) as a ruse for the families to maintain control of the company while diversifying their stake (that is, selling shares).

The Times is not one of these companies. The family is deeply committed to journalism. And it was clear, after getting to know Arthur, that the financial health of the *Times* was meaningful, but only in the pursuit of the profound—the *Times'* form of journalism. I imagine Arthur wakes up in a cold sweat frequently, fearing he could be the cousin that loses the *New York Times*.

So, the Sulzbergers, like many newspaper families, owned a minority of the equity, 18 percent, but controlled ten of fifteen seats on the board. That meant that agitators like me had to swing a whole bunch of family friends and members to the wild side. After sharing our ideas about digital and capital allocation, we continued to meet with shareholders to gauge support. Annual meetings are like elections, and shareholders—in this case the Class A shareholders—get to vote for who will represent them on the board. Most shareholders we met with were fed up and felt the Times leadership had mismanaged the company. Everything indicated the company was ripe for change.

The following week, the Times' CEO, Janet Robinson, and direc-tor Bill Kennard asked to meet with Phil, without me, to see if we could come to a settlement. This meant they knew they were going to lose at the shareholder meeting. I felt Phil should demand all four seats we had nominated directors for. But Phil said we should demonstrate some good faith and settle for two. This was a mistake: we needed several voices to break through the cacophony of the board's thoughtful comments, while not holding Arthur or Janet accountable for any real leadership.

The Times Company agreed immediately, with one condition: I wouldn't be one of the two (see above: visceral dislike). Phil recognized I had skin in the game and would not be co-opted by quarterly dinners with Nick Kristoff and Thomas Friedman, along with $200,000 in board fees (stipend and options). Instead, I would continue to push for change. So, Phil demanded I get one of the seats, and they acquiesced.

At the annual meeting in April 2008, Jim Kohlberg and I were elected to the board in an unexceptional shareholder meeting. After the meeting, Arthur asked to speak to me alone. He took me into a room and asked who the photographer was I had brought with me. I hadn't brought anybody with me. Not once, but two more times, over the next hour, he pulled me into a room and demanded that "this time" I tell him who the photographer was. "Again, Arthur," I replied with increasing irritation, "I have no fucking idea. Don't ask me again." I don't know if Arthur sees dead people or was so stressed about having an uninvited guest shoved into his boardroom that he was hallucinating. There was no photographer.

And so, our relationship began with a petty sideshow reflecting our mutual distrust and disdain. He viewed me as an unwashed mongrel, in over his head, who had no license to be on the board of the Gray Lady. I viewed him as a silly rich kid with poor business judgment. Over the next couple years, we would prove each other right.

Arthur lived and breathed the *Times*. His DNA was gray and wrapped in a blue plastic bag. It was hard even to imagine Arthur outside the building. I once saw him at a conference in Germany, and it was like seeing a giraffe on the 6 subway line—it just didn't fit.

As you might guess, I didn't have any luck convincing the board to dump the CEO, Janet Robinson, and replace her with Eric Schmidt, a man with a deep understanding of the intersection between technology and media. I was basically laughed out of the room. No one wanted to take on the CEO and Arthur. And since I was a newcomer with no credibility, it was easy to quash the idea.

This was several years before a tech CEO took over an ailing newspaper. In 2013, Jeff Bezos bought the *Washington Post*. That had the salutary effect of eliminating the quarterly freak-outs, when the paper would unveil sinking numbers to investors, soon followed by the inevitable bloodletting in the newsroom. More than providing financial ballast, Bezos turned the *Post* toward the web with a vengeance. Its online traffic doubled in three years, leapfrogging the *Times*. And the *Post* developed a content management system that it's now leasing to other news outlets. According to the *Columbia Journalism Review*, this CMS could generate $100 million a year.[24] *WaPo* is benefitting from the same blessing as Amazon: cheap capital and the confidence to invest it aggressively, and deftly, for the long term, as if they were eighteen again.

My fellow directors at the Times Company had no stomach for this type of agita. It was a lot easier, they concluded long before I came, to confront the online challenge by acquiring an online player and extending its model to the web.

About.com

In 2005, the New York Times Company purchased About.com—a growing set of sites, hundreds of them, that provided readers with specialty information about everything from pruning trees to prostate therapies.[25] It was what was known as a "content farm." The success formula for content farms was to engineer the sites with one overriding goal: leverage user-generated content that was optimized on Google, appear on the first page of Google search results, generating traffic, and thus sell advertising.

It's not fair to say that the *Times* wasn't an innovator. It was, and it became a leading site, with arresting graphics, data features, and video. But much of the online growth at the *Times* was as a collection of mediocre content engineered to capture clicks on Google (through About.com). Similar to birds in Africa that sit on a rhino's ass all day long eating mites and ticks, the *Times* was riding on the back of a titan, one of the Four. The folks at the *Times* didn't suspect it, but living on Google's search algorithm is a tenuous existence. It takes just one flick of the rhino's tail to knock the scavenging bird off.

The *Times* paid $400 million for About.com and, as the About sites harvested billions of clicks from Google searches, the purchase briefly looked deft. By the time I was aboard, the market value for About had risen to about $1 billion. About.com was hot property.

I lobbied to sell, or spin it by taking it public. Naturally, people in the About group thought this was a fabulous idea. They were sick of propping up an analog company and hungry for internet equity and respect. At one point, I made a serious faux pas: at a meeting where About's senior management was present, I suggested selling the company or taking it public. This was irresponsible on my part. It was like screaming to a room full of seven-year-old boys, "Who wants to go to Monster Jam?" when you're not sure you can get tickets.

However, Janet and Arthur didn't want to lose their online cred. They were busy using About as digital earrings to accessorize an analog outfit. It showed investors and the board (minus me) that the *Times* did have a digital strategy, one that was bringing in revenue and was poised to grow. They weren't closing their eyes to the future, they told themselves, but embracing it. Digital was bringing in only 12 percent of company revenue. Selling About would cause that number to shrink, and we might look like a newspaper company.

Meanwhile, I was pushing at board meetings for the company to shut off Google's access to *Times* content. I could see that Google's search engine already was destroying shareholder value. If left unchecked, it would slowly and methodically asphyxiate us. Everybody else believed it was the electricity of the internet age and that the relationship was symbiotic, as in exchange for our content, we got traffic from Google.

I remember one board meeting in particular. A *Times* reporter had been kidnapped in Afghanistan, and was later rescued by British Commandos. During the operation, one brave soldier was killed. The commander of the squadron wrote a moving letter to Arthur explaining why the heavy price was worth paying—why journalism matters. Arthur read the full letter to the board, pausing regularly to let us reflect before continuing. Journalism, sacrifice, deference, standing, geopolitics, ceremony. This was the giraffe on the plains of the Sudan woodlands feeding on the intermittent vegetation of edaphic grassland and acacia. Arthur was in his element.

Meanwhile, as we relished in the importance of, and sacrifices made for, journalism, Google crawlers entered our basement and scraped all our content from our servers as *New York Times* directors dined seventeen floors above in the seventh tallest building in America.

Google not only was crawling our content for free, it also was slicing and dicing that content for its users. When people were looking for a hotel in Paris, for example, Google would link to a *New York Times* travel article on Paris. But at the top of the page it would place Google's own ad for the Four Seasons Hotel. The argument was that this arrangement brought traffic to the *Times*. It could sell these eyeballs to advertisers, who would buy banner ads. It sounded good, but it was whistling through the graveyard.

Here's the rub: as it was handling those searches, Google also was learning—better than the *Times* itself—exactly what the paper's readers wanted and were likely to want in the future. And that meant Google could target those *Times* readers with far greater precision and make more money from each ad. As much as ten times more. That meant we were exchanging dollars for dimes. We should be running our own ads on our sites. What idiots we were.

Our sales team was average, and the business model was dying. The one thing we still had of value was our content—and the professionals who generate it. Yet, instead of making that content scarce—shutting off and suing any digital platform that repurposed our content—we decided that we should try to drive more traffic by prostituting our content . . . everywhere. This was the equivalent of Hermès deciding to distribute Birkin bags through walmart.com so hermes.com could get more traffic. We committed one of the great missteps in modern business history. Overnight we took a luxury brand, spread it through sewer-like distribution, and let the sewer owner charge less for it than we were charging in our own store, through subscriptions.

I was resolute, armed with data, and I represented the largest shareholder. I fantasized that one day there would be case studies about how an angry professor helped the Gray Lady, and journalism as a bonus. I made the case to the board that we needed to shut off Google's crawlers and create a global consortium of premium content. And in the hour that followed, there actually was semi-serious debate. It featured a group of mostly middle-aged, highly pedigreed gridiron greats who didn't know a fucking thing about technology. To her credit, Janet took the suggestion seriously and said management would evaluate my proposal.

A few weeks later the board received a thoughtful memo that said the *New York Times* should not shut off the search engine, as the paper couldn't risk angering Google, because About.com depended on Google for traffic. If we turned off Google, Google might counter by tweaking its algorithm, relegating About to search engine purgatory.

This, in a nutshell, is the problem with conglomerates—and the Innovator's Dilemma. The whole is often less than the sum of its parts. That was true both of the *Times* and About. In a sense, we and Google were both using each other. Google used our content to attract billions of clicks for its ads, and we used their search algorithm to drive traffic to About. However, Google had far greater power. It ruled like a lord over a crucial swath of the internet. We were the equivalent of tenant farmers on that turf. Our fate was determined from the start.

It took a while, but in February 2011, Google finally tired of the antics of content farms, including About.com, and it swatted them away. The search giant performed a "Panda algorithm update," which exiled much of the content farm traffic, and business, into Outer Oblivion. With just one tweak, Google pummeled the *Times*, diverting millions in online revenue to other sites and cutting About's value dramatically. It appears that, unlike us, Google was making business decisions based on the long-term value of the company, unafraid of our reaction. About was worth $1 billion before the update, and less than half of that the next day. A year later, the *Times* unloaded the content farm for $300 million, 25 percent less than what it had paid for it. I'd venture that "angering" About.com's parent, the Times Company, wasn't a factor in Google's decision to do what was best for Google shareholders long term.

God can give advice, influence, and, when needed, control. But, as Greek mythology teaches us over and over, sleeping with gods never ends well.

My tenure at the Times Company was not a success (understatement). My suggestions changed little. The company did sell noncore assets and decided to cancel the dividend in 2009. However, in September 2013, it reinstated the dividend—signaling that the board was firmly controlled by the family. As the Great Recession pulverized the company's ad revenue and the stock dropped, Phil Falcone decided to limit his losses by selling stock. His ownership stake was the only thing keeping me on the board. And when that started to shrink, I got the word from a couple directors that I was gone. After Arthur left me a voicemail, asking me to call him, I resigned.

I had turned $600 million, of other people's money, into $350 million. As part of our board compensation, we were granted options. Mine were worth around $10,000 or $15,000. I just needed to fill out some paperwork. I decided not to; I didn't deserve it.

Enter the New God

But God is much more: omniscient, omnipotent, and immortal. And of those three, Google is only the first—kinda. If Apple has managed to achieve a degree of immortality by converting itself into a luxury goods company, Google has accomplished the opposite: it has made itself into a *public utility*. It is ubiquitous, increasingly invisible in everyday use, and like Coke, Xerox, and Wite-Out before, it increasingly needs to reinforce the legality of its brand name for fear it will become a verb. Its market dominance is so great that it's in perpetual risk of antitrust suits at home and abroad. The EU seems to have a particular animus toward the company, filing four formal charges since 2015. The European Commission has accused Google of unfair advantage over ad competitors.[26] With a 90 percent share of EU search, and headquarters not in the EU, Google is a—rightfully— attractive target for people who are charged with policing the market.

Google divinely replied to a recent statement of objections: "We believe that our innovations and product improvements have increased choice for European consumers and promote competition."[27]

So, despite its enormous market dominance—the greatest of any of the Four—Google is also uniquely vulnerable. Perhaps that's why, of the Four, Google seems the most retiring, the most likely to remove itself from the limelight. "Gods don't take curtain calls," John Updike famously wrote of Ted Williams's refusal to come out of the dugout to acknowledge the crowd after his last at bat. Lately, Google seems to prefer to keep its cap low over its eyes, rather than doff it.

The genius of Google was there from day one, in September 1998, when Stanford students Sergey Brin and Larry Page designed a new web tool, called a search engine, that could skip across the internet in search of keywords. But the crucial step was the hiring as CEO of Eric Schmidt, a scientist turned businessman who had paid his dues at Sun Microsystems and Novell. Both of those firms had taken on Microsoft—and lost. Schmidt swore that would never happen again. Schmidt possessed a key attribute among great business leaders—an enormous chip on his shoulder—and, as Bill Gates became his great white whale, Schmidt turned his obsession into a strategy . . . and Google his *Pequod*, targeted to harpoon Moby-Dick.

It's easy to forget now that until Google came along, Microsoft had never really been defeated—in fact, it was considered the original horseman. Hundreds of companies had tried—even Netscape, with one of the most original products in tech history—and died. Microsoft is resurgent, demonstrating elephants can dance.

Google may have had just one product (that made money), but it was world changing, and the company did everything right. The goofy name and simple homepage, the honest search, uninfluenced by advertisers, the apparent lack of interest in moving into other markets, and the likable founders all conspired to make Google appealing to everyday users and apparently unthreatening to potential competitors (such as the *New York Times*) until it was too late. Google only reinforced this with nice Summer of Love philosophical statements such as "Do No Evil" and images of employees sleeping in their cubicles with their dogs.

But behind the curtain, Google was undertaking one of the most ambitious strategies in business history: to organize *all* of the world's information. In particular, to capture and control every cache of productive information that currently existed on, or could be ported to, the web. And with absolute single-mindedness, the company has done just that. It began with the stuff already on the web—it couldn't own that, but it could become the gatekeeper to it. After that, it went after every location (Google Maps), astronomical information (Google Sky), and geography (Google Earth and Google Ocean). Then it set out to capture the contents of every out-of-print book (the Google Library Project) and work of journalism (Google News).

With the insidious nature of search, Google's absorption of all the world's information took place in the open—and potential victims didn't seem to notice until it was too late. As a result, Google's control of knowledge is now so complete, and the barriers to entry by competitors so great (look at the marginal success of Microsoft's Bing), that the firm might maintain control for years.

Every company on the planet envies Google's position at the epicenter of the digital world. But the reality is less happy. Leave aside the likelihood that once the company becomes old news, Congress and the Justice Department might just decide the search engine is a public utility and regulate the firm as such.

Google is a long way from that fate—but notice that it too is basically a one-trick (and one trick only) pony. There is search (YouTube is a search engine) and there is . . . well, Android—but that's an industry smartphone standard, devised by Schmidt to counter the iPhone, and its biggest players are other companies. All of the other stuff—autonomous vehicles, drones—is just chaff, designed to keep customers and, even more so, employees pumped up. To date their contribution is less than Microsoft's fading Internet Explorer.

There are other parallels between Google and Microsoft. Microsoft at its peak was notorious for having the most insufferable asshole employees in American business. They were arrogant, smug, and totally convinced—in a classic high-tech industry mistake— that what was also luck, timing, and success was, in fact, genius. Then, when Microsoft went public, and longtime employees began to vest their stock options, they left by the thousands to pursue that genius—to very mixed results.

Finally, when the SEC and the Justice Department came calling, and Microsoft continued to crush one exciting young company after another, it suddenly became embarrassing to admit you worked for the Evil Empire. The result was that Microsoft suffered a massive loss of intellectual capital as old talent left and young talent no longer wanted to work there. Suddenly, even when Microsoft had a good product idea, it no longer seemed to be able to execute on it. It was as if the brain was willing, but the arms and legs no longer worked. Even Bill Gates took off to save the world.

Google isn't Microsoft—yet. The search firm still boasts the greatest assembly of IQ in history. Google employees don't just know they are smarter than anyone else—they are. The company famously expects employees to devote 10 percent of their workweek to coming up with new ideas—so wouldn't you expect to see a lot more interesting stuff coming from 60,000 geniuses?

Ultimately, however, it may not matter. The internet isn't going anywhere, and Google will likely continue to grow—accelerate, more likely—in its core business. Our quest for knowledge can never be sated. And Google has a monopoly on prayer, when your gaze is turned downward.

Chapter 6

Lie to Me

STEALING IS A CORE COMPETENCE of high-growth tech firms. We don't like to believe this, as entrepreneurs hold a special, exalted place in American culture. They are the spirited mavericks tilting at the windmills of giant, established corporations, T-shirted Prometheuses who bring the fire of new technologies to humankind. The truth is less romantic.

Horsemen, of course, don't start out as globally dominant megalodons. They begin as ideas, as someone's garage or dorm-room project. Their path looks obvious and even inevitable in hindsight, but it's almost always an improvisational series of actions and reactions. As with professional athletes, we tend to focus on the stories of the few who make it—and not the thousands who never get past the minor leagues. Meanwhile, the powerful, deep-pocketed company that emerges looks very little like the scrappy upstart that broke out of the garage years—especially after the corporate PR department gets done rewriting the founding myth. This transformation occurs despite how much founders fight to maintain the youthful energy of the start-up phase.

But change is inevitable, in part because the marketplace is always changing, so companies must adapt or die, but also because young companies with nothing to lose can (and do) get away with the deception, thievery, and outright falsehoods that are unavailable to companies with reputations, markets, and assets to protect. Not to mention the Justice Department doesn't care about little companies until they get big. When history is written, by the winners, terms such as *inspired by* and *benchmarked* replace less savory terms.

The sins of the horsemen fall into one of two types of cons. The first is taking—which often means *stealing* IP from other companies and repurposing it for profit, only to viciously protect that IP once they've amassed a lot of it. The second is profiting from assets built by someone else in a manner unavailable to the originator. The first means that the future horsemen don't have to depend upon their native ingenuity to come up with innovative ideas—and throwing lawyers at those who try the same thing to them means they won't be victims, too. The second is a reminder that the so-called first-mover advantage is usually not an advantage. Industry pioneers often end up with arrows in their backs—while the horsemen, arriving later (Facebook after Myspace, Apple after the first PC builders, Google after the early search engines, Amazon after the first online retailers), get to feed off the carcasses of their predecessors by learning from their mistakes, buying their assets, and taking their customers.

Con #1: Steal and Protect

Great companies often rely on some sort of lie or IP theft to accrue value at a speed and scale previously unimaginable, and the Four are no different. Most horsemen have fostered a falsehood that cons other firms, or the government, into a subsidy or transfer of value that dramatically shifts the balance of power to them. (Just watch Tesla over the next few years as it fights for government subsidies for solar- and electric-powered cars.) When they emerge as a horseman, however, they are suddenly outraged at this sort of behavior and seek to protect their gains.

This dynamic can be seen even more starkly with countries. In the geopolitical context, there is only one horseman, the United States of America, and its history demonstrates this dynamic. In the period immediately following the Revolution, the United States was a scrappy start-up, with plenty of opportunities but little capability to exploit them. In Europe, industrial innovation—the Industrial Revolution—was flowering during a period of relative peace, and American manufacturers just couldn't compete. In particular, the important textile industry was dominated by British weavers, using advanced looms (whose designs they'd stolen from the French) and related technologies. Britain sought to protect this industry with laws barring export of equipment, plans for equipment, or even the artisans who built and operated them.

So, the Americans stole it. Secretary of the Treasury Alexander Hamilton issued a report calling for the procurement of European industrial technology through "proper provision and due pains"— even as it blithely acknowledged that British law prohibited such export.[1] The Treasury offered bounties to European artisans willing to come to the United States, in direct violation of emigration laws in their home countries. U.S. patent law was modified in 1793 to limit patent protection to U.S. citizens, thus depriving European owners of this intellectual property any legal recourse against this theft.

From these seeds, America's industrial might grew quickly. The town of Lowell, Massachusetts, known as the cradle of the American Industrial Revolution, was built by the corporate descendants of Francis Cabot Lowell, who had years earlier toured British textile plants as a curious customer (which was true, if incomplete) and memorized their design and layout. Upon his return to the United States he founded the Boston Manufacturing Company and built America's first factory—and, in a nice connection with our modern tech industry, conducted the country's first IPO.[2] The thievery gave rise to a multibillion-dollar industry: consulting. The United States has the best consulting firms in the world—theft is in our DNA.

Today, the United States is the industrial behemoth, with its own technological advantages and markets to protect. And while we celebrate Alexander Hamilton on Broadway, our laws repudiate his casual attitude toward intellectual property. The United States is now the great proponent of patent and trademark protections, and you can't go wrong, as a U.S. politician, criticizing China for stealing U.S. technology. And not without cause, as China, eager to achieve horseman status on the world stage, is sending its own Francis Lowells over, in person and through cyberspace, to grab whatever can shorten the path to prosperity. Meanwhile, after decades of stealing the world's patents, China now feels strong enough in IP that it now has seen the light and is becoming a vocal advocate of patent law.

Perhaps the most famous "theft" in tech history is at the root of

Apple, when Steve Jobs turned Xerox's unfulfilled vision of a mouse-driven, graphical desktop into the industry-changing Macintosh.[3]

Like Lowell and his contemporaries, who improved upon British designs and powered them with the vast resources and growing population of the young United States, Jobs saw Xerox's GUI had the potential to explode the PC market beyond what even his massively successful Apple II had achieved. The GUI could create, as Apple famously put it, "the computer for the rest of us."[4] This was something Xerox was never going to do, was never capable—institutionally, strategically, philosophically—of doing.

So, Apple merely takes innovations developed elsewhere and uses its better "marketing." Sort of. It's certainly the case that Apple has bought or licensed many of the technological underpinnings of its current leadership position, from Xerox's GUI to Synaptics' touch screens to P.A. Semi's power-efficient chips. The point is not that young companies just "steal" things to become great, but that they see value where others don't, or are able to extract value where others can't. And they do so by whatever means necessary.

Con #2: Not Stealing, Just Borrowing

Another way the Four cheat is by borrowing your information, only to sell it back to you. Google is a good example.

Google was founded based on mathematical insights about the structure of the web and the nature of search, but it became a horseman based on the founders' (and Eric Schmidt's) insight that information could be given away free with one hand, while made very, very profitable with the other. Marissa Mayer, then an executive at Google, sat before Congress and told a bunch of mostly white, mostly old, mostly men that newspapers and magazines had a natural obligation to let information be crawlable, sliceable, query-able, and searchable . . . by Google.[5] Stories provided by Google News, she said, "are sorted without regard to political viewpoint or ideology, and users can choose from a wide variety of perspectives on any given story."[6] A thousand flowers would bloom in the ether, she implied; we'd maintain the country's DNA of innovation, and inner city kids would get their book reports done. It was similar to PBS hauling out Big Bird when it wanted its subsidy renewed. Who wants to kill Big Bird?

Indeed, Mayer testified, Google provided "a valuable free service to online newspapers specifically by sending interested readers to their sites."[7] She sounded disappointed that the *New York Times* and the *Chicago Tribune* weren't thanking Google for all it had done for them. Perhaps this was because Google's "valuable free service" was, in fact, rapidly gutting the advertising base of the American media—and rerouting all those revenues to Google.

Never fear, Mayer told Congress, Google has a valuable, although not free, service for that, too. Publishers, who were increasingly dependent on Google to source their traffic, could join Google Ad-Sense, which "helps publishers generate revenue from their content."[8]

The reality, of course, was that by the 2016 election, information had become polarized by algorithms that determine your "political viewpoint and ideology" in a millisecond.[9] In the period after Mayer delivered her testimony, news publishers—who before Google had no need for "help" generating revenue—disappeared at an alarming rate. Meanwhile, Google hoovered up information—about us, about our habits, about our world—turning its algorithms loose on that information to bring us more "valuable free services."

Both Facebook and Google stated, earlier in the decade, that they would not share information across silos (Facebook to Instagram, Google to Gmail to YouTube to DoubleClick). However, both lied and have quietly changed their privacy policies, requiring a specific request to opt out if you don't want them to cross-reference your movements and activity against location and searches. There is no evidence of any intent beyond the data being used for better targeting. Creepy and relevance are strongly correlated in the world of digital marketing. To date, consumers and advertisers have voted with their actions and expressed that creepy is a price worth paying for the relevance.

Information's Price Tag

The hacker credo "Information wants to be free" set the stage for the second golden age of the internet. The origin of the phrase is worth reviewing. First proposed by Stewart Brand, the founder of the *Whole Earth Catalog*, at the 1984 Hackers Conference, his formulation was:

> On the one hand information wants to be expensive, because it's so valuable. The right information in the right place just changes your life. On the other hand, information wants to be free, because the cost of getting it out is getting lower and lower all the time. So, you have these two fighting against each other.[10]

Information, like the rest of us, desperately wants to be attractive, unique, and well paid—*really* well paid. Information wants to be expensive. The most successful media company in America, other than Google and Facebook, is Bloomberg. Michael Bloomberg never fell for it—giving information away. He mixed other people's information with proprietary data, added a layer of intelligence and—here's the trick—made it scarce. It was expensive and had its own vertical distribution (storefronts) in the form of Bloomberg terminals. If you want breaking business news that might impact the price of a stock in your portfolio, you sign up with Bloomberg, get a terminal installed in your office, and soon the screen is rolling with an endless flow of news and financial data.

The "information wants to be expensive" part of Brand's quote seems to have been erased, like Trotsky from photos, by firms looking for content for free. Indeed, it was the tension between the two that Brand was interested in, and it was in that tension where he foresaw innovation. Google (and Facebook in a different context) has mastered that tension. It takes advantage of the declining costs of distribution by giving its users access to a world of previously expensive information, then extracts billions in value by being the new gatekeeper.

Facebook too has leveraged the tension between information's ever-lower costs and its persistently high value. Its jujitsu move is even more dramatic than Google's. Facebook gets its users to create the content, then it sells that content to advertisers so they can advertise to the users who made it. It's not "stealing" our baby pictures and political rants, but it is extracting billions of dollars from them using technology and innovation unavailable to us as individuals. That's world-class "borrowing."

Facebook built its foundation on a second lie, repeated thousands of times in early meetings between Facebook's army of sales reps and the world's largest consumer brands: "Build big communities and you will own them." Hundreds of brands invested hundreds of millions on Facebook to aggregate enormous branded communities hosted by Facebook. And by urging consumers to "like" their brands, they gave Facebook an inordinate amount of free advertising. After brands built this expensive house, and were ready to move in, Facebook barked, "Just kidding, those fans aren't really yours; you need to rent them." The organic reach of a brand's content—percentage of posts from a brand received in a fan's feed—fell from 100 percent to single digits. Now, if a brand wants to reach its community, it must advertise on—that is, pay—Facebook. This is similar to building a house and having the county inspector show up as you're putting on the finishing touches. As she changes the locks she informs you, "You have to rent this from us."

A mess of big companies thought they were going to be Facebook owners and ended up Facebook renters. Nike paid Facebook to build its community, but now less than 2 percent of Nike's posts reach that community—unless, that is, they advertise on Facebook.[11] If Nike doesn't like it, tough shit, they can go cry to the community on the world's other two-billion-member social network . . . oh wait. Similar to someone dating a person much hotter than them, brands complained and took the abuse.

The Key to a Great Con

It's pretty clear where Amazon is headed: 1) Take over the retail and media sectors, globally; and 2) Replace the delivery of all these products (goodbye UPS, FedEx, and DHL) with its own planes, drones, and autonomous vehicles. Sure, they'll continue to hit speed bumps. However, their culture of innovation and access to infinite capital will roll right over them. Does anyone believe that any country (except perhaps China, protecting its own online retailer Alibaba) can resist?

As Paul Newman explained in *The Sting*, the key to a great con is that the victim never realizes he was conned—indeed, he believes he is about to be a big winner right up until the last moment. Newspapers still feel that the future happened *to* them versus what really happened: they were Googled, hard. And where Google didn't molest them, their own stupidity—not buying eBay when it was offered to them on a silver platter, not snapping up Craigslist when it was still a start-up, and keeping their top talent on the print side instead of moving them to the web—doomed them. Had they made the right decision on just half of the opportunities presented to them by cyberspace, most would still be around.

The rest of the Four have similarly pulled the wool over the eyes of their victims. Brands eagerly pumped money into building Facebook communities before realizing they weren't their own. Sellers are quick to join Amazon, believing the platform provides them access to a new swath of customers, but then find themselves in competition with Amazon itself. Even Xerox thought it was getting a lucrative piece (100,000 shares) of Apple, one of the world's hottest tech companies, for simply letting Steve Jobs look behind "its kimono."[12] You could say these wounds were self-inflicted.

Aspirational horsemen always show a willingness to go to market in ways unavailable to their old-guard competition. Uber, for example, operates in blatant contravention of the law in many, perhaps just about all, of its markets. It has been banned in Germany; Uber drivers are fined in France (but Uber pays the fines);[13] and various U.S. jurisdictions have ordered Uber to cease operations.[14] And yet, investors—including governments—are lining up to hand the company billions. Why? Because they sense that, in the end, the law will give way before Uber does; Uber is inevitable. And they are probably right. There are laws, and there are innovators. Good money is on the innovators.

Uber not only evades the regulations traditionally applicable to car-for-hire services; it also evades labor law by posing as an app that links independent drivers—a posture that nobody seriously believes. Yet despite all of this, Uber continues to sign up drivers and riders at a furious pace—myself included—because its basic service and simple app are vastly superior to the coddled, protected taxi model. Uber has recognized that if an industry is broken enough, consumers will conspire to violate the law in favor of a far preferable service. And, in the long run, do you really think Congress is going to fight both Wall Street and millions of consumers?

Amazon has also effectively conspired with half a billion consumers to use algorithms to starch the margin brands used to garner and deliver those savings to their ally, consumers. A retailer leveraging its power to grow a higher-margin private label is not new. We've just never seen anybody this good at it. Just as U.S. allies were "shocked" we were listening in on world leaders' phone calls, they all knew we spied on each other. What pissed them off is how much better we, the United States, are at it. This alliance between Amazon, consumer, and algorithms gives consumers enormous value, and Amazon's resultant (blistering) growth garners hundreds of billions in shareholder value for employees and investors. As consumers we benefit enormously from a relationship with the most powerful allies you could ever have on your side. As citizens, wage earners, and competitors, we know we are being abused but just can't break up with the hot girl.

There is a justice system, but it isn't blind. It's good to be as rich as one of the Four when caught red-handed. Facebook assured EU regulators seeking approval of the acquisition of WhatsApp that it would be impossible for the two entities to share data in the short term. This promise assuaged regulators' concerns over privacy, and the acquisition was approved. Spoiler alert: Facebook figured out how data could jump silos . . . pretty fast. So, feeling lied to, the EU fined Facebook 110 million euros. This is tantamount to getting a $10 parking ticket for not feeding a meter that costs $100 every fifteen minutes. The smart choice: break the law.

Chapter 7

Business and the Body

IN THEIR BESTSELLING BOOKS, Ben Horowitz, Peter Thiel, Eric Schmidt, Salim Ismaiel, and others argue that extraordinary business success requires scaling at low cost, achieved by leveraging cloud computing, virtualization, and network effects to achieve a 10x productivity improvement over the competition.[1] But that explanation ignores a deeper dimension that has nothing to do with technology. From the perspective of evolutionary psychology, all successful businesses appeal to one of three areas of the body—the *brain*, the *heart*, or the *genitals*. Each is tasked with a different aspect of survival. For anyone leading a company, knowing which realm you play in—that is, which organ you inspire—dictates business strategy and outcomes.

Gray Matter

The brain is a calculating, rational thing. To do its job, the brain weighs costs and benefits, balancing tradeoffs by the millisecond. In the marketplace, the brain compares prices and applies the brakes with blistering speed. If it learns that Huggies diapers are 50 cents cheaper than Pampers, it enacts a complex cost-benefit analysis, including past experience with the two types of diapers—which one absorbs better?—and calculates the better choice. This translates, in the business world, to lower margins. For most businesses, the consumer's brain is the ultimate buzzkill and competitor. Lincoln was right about not being able to fool all of the people all of the time—and a lot of dead companies have regretted trying to do so. Our brains prevent us from making too many dumb decisions, at least after we screw up a few times.

A handful of companies address their customers' brains, appeal to our rational selves, and manage to win. Take Walmart: millions of consumers size up their options and shop there. As a value proposition, "more for less" has been the bomb for a long time. It's why our ancestors, when deciding which animals to stampede over cliffs, chose bison over chipmunks, even though the former presented a whole lot more risk.

Walmart runs one of the world's most efficient supply chains at an unrivaled scale. So, the retailer holds its suppliers (makers of mass-produced, commodified goods) by the balls. By squeezing them, Walmart lowers its costs, enabling it to pass the savings to consumers and expand its market share. Walmart now commands approximately 11 percent of retail in the United States.[2] Despite low margins, its sheer volume produces gargantuan profits. Walmart's customers use their brains well—arguably better than wealthier people who spend more in exchange for perceived prestige.

There is enormous shareholder value created by the winner in the fight against the brain, but it's winner-take-all. Once the brain determines the rational best choice, it's decisive, monogamous, and loyal. The poster children for competing in the battle for the brain are Walmart, Amazon, and even China (they compete on price). Most companies are not, and can never be, the low-cost leader. It's a select club that demands scale for long-term success.

But what if you're not, and do not aspire to be, the logistics king? Then your focus should migrate south, from the cold, rigid calculations of the brain to the more forgiving heart.

Bighearted

The heart is a vast market. Why? Because most of our actions, including purchases, are driven by emotion. It's easier, and more fun, than to turn to the killjoy brain for a predictable cost-benefit analysis, where the answer to "Should I buy this?" is usually "No." The heart is also powered by the greatest force in history: *love*.

We feel better when we love, nurture, and care for others. We also live longer. The Okinawa Centenarian Study examined the lives of people on the southern Japanese island, one of the world's blue zones for centenarians. The researchers found that these old folks ate a lot of beans and drank every day (good news) in moderation (bummer).[3] They also exercised daily and were social animals.[4] Finally, they loved and cared for large groups of people.[5] Recent research from the Johns Hopkins University Center on Aging and Health found that caregivers had an 18 percent lower mortality rate than noncaregivers.[6] Love keeps you alive. It's Darwinian—the species needs caregivers to skirt extinction.

The heart may be irrational, but as a business strategy, targeting the heart is a shrewd and sober strategy. In fact, the explosion of consumer marketing that followed World War II was targeted, almost exclusively, at the heart. The brands, slogans, and jingles were engineered to latch onto what mattered most to consumers—what they loved. The heart was the relentless focus of the real-life Don Drapers of Madison Avenue. Thus, J.M. Smucker convinced people the love they felt for their children was directly correlated to the peanut butter they chose: "Choosy moms choose Jif." Love is also the key to seasonal promotion, from Christmas to all the Hallmark holidays: "Show Mom how much you love her." A diamond engagement ring that cost three months' salary was "forever." Forever for 50 percent of us, that is.

For a marketer, each string tugging at the consumer's heart translates to margin. There's (among others) beauty, patriotism, friendship, masculinity, devotion, and above all, love. These are values you can't put a price on—but marketers do. And that provides the markets of the heart with a cushion. Even if their competitors gain an edge (like logistics or value), they can survive, and even thrive, as long as they continue to connect with their customers' irrational hearts.

If this all sounds superficial, it is. That's the nature of passion—and the heart is one of the few forces that can override the decisions of the head.

The digital age, with its transparency and innovation, has declared war on the heart. Search engines and user reviews are adding a level of transparency that's starching much of the emotion from purchase decisions. Google and Amazon have signaled the end of the brand era, as consumers are less apt to defer to emotion when

YOY PERFORMANCE OF TOP CPG BRANDS
2014 – 2015

LOST SHARE   **DECLINED IN SALES**

"A Tough Road to Growth: The 2015 Mid-Year Review: How the Top 100 CPG Brands Performed." Catalina Marketing.

god (Google) or his cousin (Amazon) tell you to not be stupid and buy Amazon-branded batteries (a third of all batteries sold on the internet) vs. Duracell. The consumer packaged goods (CPG) sector, which may be the largest consumer sector in the world, was built on the heart-to-purchase relationship. In 2015, 90 percent of CPG brands lost share, and two-thirds experienced revenue declines.

What's a brand without scale to do? Either die or move further south to an even less rational organ.

Erogenously Yours

With appeals to the heart increasingly difficult, brands that appeal to the genitals are thriving. These organs drive desire and the relentless instinct to procreate. After survival, nothing rings louder in our ears than sex. Fortunately for marketers, sex and mating rituals overwhelm the brain's killjoy warnings about risk and expense. Just ask any sixteen-year-old—or a fifty-year-old shopping for a sports car.

When we're in a mating frame of mind, we seek brain silencers. We drink. We take drugs. We dim the lights, as light is a tool for the brain, and turn up the music. A study of men and women who engaged in an uncommitted sexual encounter that included penetrative sex showed 71 percent were drunk at the time.[7] These people purposefully, chemically switched off their brains, creating "compulsory carelessness."[8] If you wonder, the next morning, "What was I thinking?" you weren't. Few drunk people pull out their smartphone to compare the cost of a Grey Goose and soda at nearby bars, as they do when shopping for a Nespresso coffeemaker.

We're irrational and generous when under the influence. The combination of alcohol and the pursuit of youth leaves us swimming in hormones and desire. We're very in the moment. Luxury brands have understood this for centuries. They bypass cognition and love, tying their business to sex and the broader and pleasure-packed ecosystem of mating rituals. Men have been driven, since humanity's caveman days, to spread their seed to the four corners of the Earth. Men strut power and wealth, attempting to signal to females (or, in some cases, other men) we'll be good providers; our progeny are more likely to survive. The Panerai watch you're wearing signals to potential partners that if they mate with you, their offspring is more likely to survive than if they mate with someone wearing a Swatch.

By comparison, women's evolutionary role is to attract as many suitors as possible, so as to select the most promising—strongest, fastest, smartest—mate. To this end, women will contort into a $1,085 pair of ergonomically impossible Christian Louboutin platform shoes rather than wear comfortable twenty-dollar flats.

These decisions, if you can call them that, cast the consumer and provider into a symbiotic relationship. The consumer spends more because the act of spending itself communicates taste, wealth and privilege, and desire. The company, naturally, is dedicated to the same proposition, but in reverse, by providing consumers with the tools of that communication. It knows that if its products work as mating brands—the market equivalent of peacock feathers—then higher margins and profits will follow, frustrating the brain and making the heart jealous. Whether it's Christian Dior, Louis Vuitton, Tiffany, or Tesla, luxury is irrational, which makes it the best business in the world. In 2016 Estée Lauder was worth more than the world's largest communications firm, WPP.[9] Richemont, owner of Cartier and Van Cleef & Arpels, was worth more than T-Mobile.[10] LVMH commands more value than Goldman Sachs.[11]

The Horsemen and the Body Framework

The body framework—brain, heart, and genitals—bears directly on the extraordinary success of the Four Horsemen.

Consider Google. It speaks to the brain, and supplements it, scaling up our long-term memory to an almost infinite degree. It does so not only by accessing petabytes of information around the globe—but just as important, substitutes for our brain's complex and singular search "engine" (and its ability to shortcut at a fantastic speed across the dendrites of brain neurons). To that remarkable physiological ability, Google adds the brute force of ultrafast processing and high-speed broadband networking to race around the world to find, on the right server, the exact piece of information we desire. Human beings, of course, can do the same thing—but it would probably take weeks and a lot of travel to some dusty library to find the same thing. Google can do all that in less than a second—and offers to find for us the next obscure fact, and another after that. It never tires, it never gets jet lag. And it not only finds whatever we're looking for . . . but a hundred thousand other similar things we might be interested in.

Finally, and ultimately most important, we *trust* the results of Google searches—even more than our own, sometimes fitful, memories. We don't know how the Google algorithm works—but trust it to the point of betting our careers, even lives, on its answers.

Google has become the nerve center of our shared prosthetic brain. It dominates the knowledge industry the way Walmart and Amazon, respectively, rule offline and online retail. And it certainly doesn't hurt that when Google reaches into our pockets, it's mostly for pennies, nickels, and dimes. It's the antithesis of a luxury company—it's available to everyone, anywhere, whether they are rich or poor, genius or slow. We don't care how big and dominant Google has become, because our experience of it is small, intimate, and personal. And if it turns those pennies into tens of billions of revenue, and hundreds of billions in shareholder value, we aren't resentful—as long as it gives us answers and makes our brains seem smarter. It is *the* winner, and its shareholder benefit stems from the brain's winner-take-all economy. Google gives the consumer the best answer, for less, more quickly than any organization in history. The brain can't help but love Google.

If Google represents the brain, Amazon is a link between the brain and our acquisitive fingers—our hunter-gatherer instinct to acquire more stuff. At the dawn of history, better tools meant an improved and longer life. Historically, the more stuff we had, the more secure and successful we felt. We felt safer from our enemies and superior to our friends and neighbors. And who could ask for more? People dismiss Starbucks' success as simply "delivering caffeine to addicts." But caffeine is Nicorette compared to the heroin of shopping.

Facebook, by contrast, appeals to our hearts. Not in the manner that the Tide brand appeals to your maternal instincts of love, but in that it connects us with friends and family. Facebook is the world's connective tissue: a combination of our behavioral data and ad revenue that underwrites a Google-like behemoth. However, unlike Google, Facebook is all about emotion. Human beings are social creatures; we aren't built to be alone. Take us away from family and friends and, research has shown, we'll have a greater chance of depression and mental illness, and a shorter life.

Facebook's genius was not just in giving us yet another place on the web to establish our identities, but also the *tools* to enable us to enrich that presentation—and to reach out to others in our circle. It has long been known that people exist in groups of a finite and specific size. The numbers repeat themselves throughout human history, from the size of a Roman legion to the population of a medieval village . . . to our number of friends on Facebook. These numbers have a very human source: we typically have one mate (2 people), the people we consider very close friends—as the joke goes, people who will help you move a body (6 people), and the number of people we can work with efficiently as a team (12), up to the number of people we recognize on sight (1,500 people). The unseen power of Facebook is that it not only deepens our connections to those groups, but by providing more powerful, multimedia lines of communication, it expands our connections to more members. This makes us happier; we feel accepted and loved.

Apple started out in the head, firmly in the tech sector's vocabulary of logistics. It boasted efficiency: "Ford spent the better part of 1903 tackling the same details you'll handle in minutes with an Apple," read a print ad. The Mac helped you "think different." But finally, Apple has migrated further down the torso. Its self-expressive, luxury brand appeals to our need for sex appeal. Only by addressing our procreative hungers could Apple exact the most irrational margins, relative to peers, in business history and become the most profitable firm in history. When I was on the board of Gateway, we operated (poorly) on 6 percent margins. Apple computers—not as powerful—garnered 28 percent. We, Gateway, had been relegated to the brain (Gateway didn't make you more attractive), where Dell had already won the (rational) scale game. We were in no-man's-land and sold for scrap. Having reached $75/share several years earlier, we sold for $1.85/share to Acer.

The lust for Apple-branded goods has given the company its cult-like status. People who belong to this cult pride themselves on their hyperrational choice to buy Apple products based on their ergonomic design, superior operating system, and resistance to viruses and hackers. Like the kids who sell them Apple products, they consider themselves "geniuses," illuminati, foot soldiers in the Apple crusade to think different and change the world. Most of all, they think it makes them *cool*.

But people outside the cult see it for what it is: a rationalization for something a lot closer to *lust*. Android users assuage their jealousy with their rational self. Buying Apple is irrational (spending $749 on a phone when you can have a similar one for $99). And they would be right. You don't camp out in front of a store waiting for the next-generation iPhone because you're making a sound decision.

Apple's marketing and promotion have never been traditionally sexy. The message is not that owning an Apple product will make you more attractive to the opposite (or same) sex. Rather—and this is common with great luxury brands—the message is that it will make you *better* than your sexual competitors: elegant, brilliant, rich, and passionate. You will be perfection: cool, shit together, listening to music in your pocket and swiping through pics of your latest trip that look professional but that you took on your phone. You'll have the ultimate earthly life. You'll feel closer to God. Or at least closer to the Jesus Christ of business, the pinnacle of success, the uncompromising genius, sexy beast Steve Jobs.

Business Growth and Biology

It would seem the Four Horsemen already have a monopoly on the key organs of the human body. So, what's left? And if there is no other great market opportunity, how do you compete with *them*?

Let's take the latter first. The current horsemen look so gigantic, rich, and dominant that it would seem impossible to attack them directly. And that's probably the case, but history suggests there are other strategies. After all, each of these companies in their day had to take on equally dominant and established corporate giants—and beat them.

For example, when Apple started out it faced several huge competitors. IBM was one of the biggest companies in the world and dominated electronics in the workplace (as the saying went, "Nobody ever got fired for buying Big Blue"). Hewlett-Packard, almost as big a company, and arguably the best-run company of all time, owned the scientific handheld and desktop calculator business. And

Digital Equipment was running neck and neck with both companies in minicomputers—and winning. How could Apple, started by two scruffy phone hackers in a garage, possibly compete with these monsters?

It did so with a combination of fearlessness, superior design, and luck. You know about the first two, but the third might surprise you. Steve Jobs knew he had a world-class product in the Apple II thanks to Woz's brilliant architecture and his own elegant design. But no corporation was going to buy his computers when they could buy inferior, but adequate, machines at a lower price and guaranteed volume delivery.

So, Jobs instead went after the individual consumer. There, he had free reign: his small competitors were stuck building hobby machines that average folks didn't trust or understand. Meanwhile, IBM was staying out of personal computers because it was fighting antitrust indictments over its mainframe computers; DEC had dismissed the idea of consumer computers, and HP—even after Woz offered Apple to Bill Hewlett—decided to focus on engineers and other professionals. Within three years of its founding, Jobs and Apple owned the personal computer market.

Then something interesting happened: those same consumers started sneaking their Apple computers into the office. It wasn't long before an insurgency was in full flower, with individual employees by the thousands using their Apple computers at work in violation of the rules put down by their employers' IT departments. That was the beginning of Apple "cool"—users felt like mavericks, corporate guerillas, fighting the Man in the MIS department. That's why, when IBM finally unleashed its PC, it destroyed the rest of the personal computer industry. But Apple, like the tiny mammal skittering under the feet of a dinosaur, survived . . . and eventually triumphed.

Google did the same thing by pretending to be small, cute, and honest with its simple homepage—even after it crushed all other search engines. Remember, Google started on Yahoo, which decided to outsource search to the little engine that could—and did: Google became a hundred times more valuable than Yahoo, which didn't see the threat. Facebook defeated the dominant Myspace by being the nice, safe alternative that wasn't overrun with sexual predators, or at least the fear of them. Facebook's roots on Ivy League college campuses made it feel more upmarket and safe: it demanded a .edu email address. The requirement to confirm, and share, one's identity created a different, more civilized decorum on Facebook.

Content on Twitter is more likely to get a hostile response than when posted on Facebook, since, similar to real life, it's easier to be an asshole anonymously. Amazon was careful never to portray bookstores as competition, even asserting that they wanted them to survive—the same way a reticulated python feels bad for the cute little mammal it suffocates and swallows whole. Similarly, as Amazon invests billions in last-mile delivery, Mr. Bezos claims Amazon has no intention of replacing UPS, DHL, or FedEx, but to "supplement" them. Yeah, that's it, Jeff and Amazon are here to help.

There is no reason to believe that these strategies—insurgency, false humility, security, and simplicity plus discounting—won't work again one day against the horsemen. Giant companies face their own challenges: they lose their best talent to more rewarding start-ups; their physical plant grows old; their empires grow so big they can no longer coordinate all their pieces; they get distracted by investigations by envious or nervous governments. The processes put in place to scale begin slowing the firm down, as managers begin believing that adhering to the guidelines is more important than making good decisions. Bezos insists that there will never be a Day 2.[12] It may seem unlikely that Amazon will one day lose its way. It will. Business mimics biology and, thus far, the mortality rate is 100 percent. The same is true of the Four. They will die. The question is not if, but when, and by whose hand?

Chapter 8

The T Algorithm

AT SOME POINT, there will be a Fifth Horseman, a company that combines a market valuation of one trillion dollars with sufficient market dominance to define its corner of the world. Or more likely, one of the Four will be replaced. Can we identify companies more likely to join this elite group?

While history may not repeat itself, it does rhyme, as Mark Twain purportedly said. Among the Four, these eight factors are prevalent: product differentiation, visionary capital, global reach, likability, vertical integration, AI, accelerant, and geography. These factors provide an algorithm, rules for what it takes to become a trillion-dollar company. In our work at L2, we use the term *T Algorithm* to help firms better allocate capital.

Here are the eight factors:

1. Product Differentiation

The key competence around building shareholder value in retail used to be *location*. People didn't have the opportunity to go much farther than the corner store. Then it was *distribution*. The railroads gave consumers the opportunity to enjoy different products produced at scale, which lowered prices and gave them brands they could depend upon.

We then moved to an era of *product*, especially in the automobile and appliance industries, largely fostered by the innovation that was a peace dividend from World War II. We got better cars, washing machines, television sets, even better apparel. The leather bomber jacket was invented in World War II, as was Silly Putty, the radar, the microwave, the transistor, and the computer. That led to the *financial* age, in which a group of companies, using cheap capital to roll up other companies into conglomerates, built the ITTs of the world. This in turn was followed by the great *brand* age of the eighties and nineties, when the key to building shareholder value was to take an average product—shoe, beer, soap—and build aspirational, intangible associations around it.

As discussed in the second chapter, we are again back to an era of *product,* as new technologies and platforms—be it Facebook or Amazon user reviews—let consumers conduct diligence across a broad array of products in a fraction of the time it used to take to shop. The ability to conduct diligence has never been easier, which reduces the need to default to brand or reputation. Now, the best product has a better chance of breaking through the clutter—whereas before, the best product without any marketing was like a tree falling in the forest. Moreover, the injection of digital "brains" into otherwise static, inanimate products has ushered in a new wave of innovation in which custom, personalizable apps can be quickly downloaded and upgraded without the need to replace the original "box."

A mattress is a mattress until you get an iPad and some basic technology. Then you can program in an "ultimate sleep number," and so can your partner. Or you can order the best mattress online, avoiding those damp warehouses called mattress stores, and have it delivered in a box and (cooler yet) watch it unfurl when dumped out of the box.

I have to take my car to the dealership to get a tune-up. My neighbor has his tune-up transmitted wirelessly into his Tesla's operating system. The engine receives an upgrade and instructions to remove the speed regulator, and the car's top speed increases from 140 to 150, remotely. Do you remember who made your landline phone, before chips and wireless set them free?

Nearly every product in the world, even products and services that appear to have been commoditized, have forged new dimensions and consumer value, enabled by cheap sensors, chip sets, the internet, networks, displays, search, social, and so on. Today, almost every link of the supply, manufacturing, and distribution chains has a new means of differentiation. All of a sudden, products driven by technology and defensible IP are the bomb.

However, don't be trapped into thinking that product differentiation is about the widget you're selling. Differentiation can occur where consumers discover the product, how they buy it, the product itself, how it's delivered, and so on. A worthwhile exercise is to map out the value chain of your product or service from the origin of the materials through its manufacture, retail, usage, and disposal... and identify where technology can add value, or remove pain, from the process/experience. You'll find that this value can affect every step—and if you happen to spot a step where it hasn't, start a new company there. Amazon is adding technology and billions to the fulfillment segment of the consumer experience that will likely create the most valuable firm in the world. Before Amazon, ordering from Williams-Sonoma meant you would pay $34.95 to get the product in a week. Now it's free in two days or less. The most mundane part of the supply chain ended up being the most valuable in the history of business.

Removal

When brainstorming for new ideas, entrepreneurs have a tendency to focus on what can be added—how to enhance the experience—instead of what can be taken away, thus making it less painful. But I'd argue that the majority of stakeholder value created over the last decade has been a function of *removal*. We, as a species, have mostly figured out what makes us happy: time with loved ones, physical and mental stimulation, substances that heighten or deaden those feelings, Netflix, and sassy church signs.

You may be tempted to think that competitive advantage in the internet age comes down to simply "more for less." After all, that's an obvious edge enjoyed by Amazon. But what about Apple? It's almost always the premium-priced brand—and though its products are typically better than the competition, they usually aren't *that* much better for the prices Apple charges. I would argue Amazon could charge as much for its products as do its brick-and-mortar competitors . . . and would still dominate the marketplace. Why? Because it's still infinitely easier to hit a couple keys on your computer to buy a book or a piece of furniture than it is to drive down to the local mall, find a parking place, walk a half mile, be overwhelmed by tons of irrelevant merchandise, and then lug your shit back to your car for the drive home. Amazon has *removed* all that friction and brings your purchases to your door for less than the cost of gas for your own car.

So, while it may seem that the value explosion brought by the technology revolution comes from the addition of new features and capabilities, its greater contribution comes from removing obstacles and time killers from our daily lives.

Friction is everywhere. For example, there is a ton of friction in transportation. That's why Uber saw an opportunity, via GPS, texting, and online payment, and removed the pain and anxiety of ordering a car, wondering "where the hell is the car," and fumbling around in the back of the car at the end of the journey trying to dig out money and pay. How many of us recently have bombed out of a taxi without paying because we've become so used to frictionless Uber? Bottom line: paying is friction, and it is disappearing. Just as hotel checkout disappeared a decade ago, check-in will be a thing of the past in another ten years. Some of the better hotels in Europe no longer require you to sign a bill after a meal. They know who you are and will charge you. Less is more.

Each of the Four has a superior product. It sounds old school, but Google really does have a superior search engine. The Apple iPhone is a better smartphone. The cleanliness of Facebook's feed—coupled with the "network effect" (the fact that everybody's on it) and a constant stream of new features—makes it a better product. Amazon redefined the shopping experience and expectations: from 1-click ordering to getting your product within two days (or in hours, soon by drone or a truck UPS used to own).

These are tangible innovations and points of product differentiation. All have been achieved through access to cheap capital set against deft technological innovation. "Product" is experiencing a renaissance, and is the first factor in the T Algorithm. If you don't have a product that is truly differentiated, you have to resort to an increasingly dull, yet expensive, tool called advertising.

2. Visionary Capital

The second competitive factor among the Four is the ability to attract cheap capital by articulating a bold vision that is easy to understand. In chapter 4 we discussed how visionary capital works for Amazon, but it is an advantage shared by three of the Four Horsemen.

Google's vision: *Organizing the world's information.* Simple, compelling, and a reason to buy the stock. Google has more money to invest in engineers than any media company in history. That lets it design more "stuff," including autonomous vehicles.

Facebook's vision: *Connecting the world.* Consider how important and generally awesome that could be. Facebook is now worth more than Walmart, and surpassed $400 billion in market value.[1] Similar to Google, it too can place more bets and offer more generous parental leave, hire buses that transport you to work, turn the roof of your office building into a park, and even pay for you to freeze your eggs so you can delay that whole procreation thing and devote yourself to a *real* contribution to the species—connecting the world.

Meanwhile, over Thanksgiving weekend 2016, Amazon captured the largest overall share of organic results for top gift items.[2] Amazon is Google's biggest customer. Is search a skill set? No doubt, Amazon is great at search, but its SEO skill would be Wayne Gretzky without a stick if it didn't throw tens of millions of cash at the issue. One in six people start their search for products using Google,[3] making it the equivalent of the second biggest (first is Amazon) retail store window in the world. Fifty-five percent start on Amazon. Take Macy's windows on Christmas and make it the size of Everest and K2—that's the size of the windows into the world that Google and Amazon search results represent in the fastest growing channel: online commerce.

Anyone can purchase a place in that window and land at the top of a Google search. When someone types in "Star Wars action figures," the retailer that has bid the most is going to top the paid listings. Amazon regularly buys that number one spot, because it has the money to do so. And Amazon can afford this at a scale that no one else can match, as it has cheaper capital. The company is playing by a different set of rules, and with a whole different deck of cards. As J.Crew chairman Mickey Drexler points out, "It's impossible to compete with a big company that doesn't want to make money."

The strength of visionary capital begets competitive strength. Why? Because you can more patiently nurture assets (invest) and place more bets on more pockets of innovation (try crazy shit that just might change the game). Of course, you ultimately have to show shareholders tangible progress against your big vision. However, if you're able to make the jump to light speed, and the market crowns you the innovator, the reward is an inflated valuation . . . and the self-fulfilling prophecy ("we're #1") that comes from cheap capital. The ultimate gift, in our digital age, is a CEO who has the storytelling talent to capture the imagination of the markets while surrounding themselves with people who can show incremental progress against that vision each day.

3. Global Reach

The third factor in the T Algorithm is the ability to *go global*. To be a truly large, meaningful company, you need a product that leaps geographic boundaries and appeals to people on a global scale. It is not just the bigger marketplace, but the diversity—not least the prospect of countercyclical markets that can ride out a downturn elsewhere in the world—that investors want, again rewarding you with cheaper capital. If you have a product that appears to have global reach, you're accessing 7 billion consumers vs. 1.4 billion in China, or 300 million in the United States or the EU.

Again, it doesn't require world domination, but rather proof that your product or service is so "digital-ish" that the normal rules of cultural friction do not apply. Uber's revenue growth in countries outside the United States has a chaser effect on the firm's valuation (its multiple of revenues), and the first dollar earned outside the United States increased the value of the firm by billions. If you want to be a horseman, your product needs to get a passport—that is, go global—before the kid starts kindergarten (five years old or less). Was this true of the Four when they started? No, except for Google. But the very presence of the horsemen has subsequently changed the rules.

Apple today defines what it means to be global: the brand has largely been accepted in every sovereign nation. Google has also done a good job—it's strong in mature markets—but it's been kicked out of China. Facebook has two-thirds of its users outside the United States[4] (though half of revenues are clocked in the United States);[5] its biggest market in terms of users is Asia,[6] which presents robust growth opportunities. Amazon is growing faster in Europe

PERCENT OF GLOBAL REVENUE OUTSIDE THE U.S.
2016

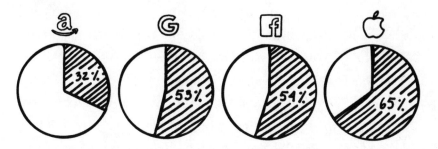

"Facebook Users in the World." Internet World Stats.
"Facebook's Average Revenue Per User as of 4th Quarter 2016, by Region (in U.S. Dollars)." Statista.
Millward, Steven. "Asia Is Now Facebook's Biggest Region." Tech in Asia.
Thomas, Daniel. "Amazon Steps Up European Expansion Plans." The Financial Times.

than in the United States.[7] It's still not as big in Asia, but it's a global company.

4. Likability

The world of commerce is regulated. Government, independent watchdog groups, and the media play a large role in a company's growth. If you are perceived as a good actor, a good citizen, caring about the country, its citizens, your workers, the people in your supply chain that get you the product, you have created a barrier against bad publicity. In the words of Silicon Valley marketer Tom Hayes, who did just that for Applied Materials, "When the news is negative, you want to be perceived as a good company to which a bad thing has happened." Image matters, a lot. Perception is a company's reality.

That makes the importance of being likeable, even cute, the fourth factor in the T Algorithm.

Bill Gates and Steve Ballmer were neither likable nor cute. In fact, the room got brighter whenever they left it. So, when Microsoft achieved a certain level of influence, district attorneys and regulators woke up one morning all over Europe and decided the easiest way to the governor's mansion, or Parliament, was to go after the Wizards of Redmond. The less likeable a company, the sooner the regulatory intervention—antitrust, antiprivacy—as questions about its supply chain or any manner of rational concerns are irrationally selected and applied. We are under the general illusion this process is more thoughtfully examined and based on some sort of equity or the law. Not true: the law decides the outcome, but the rush, or lack thereof, to drag companies into court is subjective. And that opinion is largely based on how nice or chastened the company is perceived to be.

If you remember, the feds went after Intel Corp. at the same time they went after Microsoft—both for monopolistic behavior. Intel's CEO Andrew Grove was one of the scariest figures in American industry. Yet, when the feds came calling, Andy did one of the biggest mea culpas in business history. He all but flung himself on the mercy of the SEC . . . and was forgiven. Meanwhile, Bill Gates, a far less intimidating figure, decided to play tough with the feds— and ten years later was viewed as having fallen from grace.

Google is a whole lot cuter than Microsoft. And Sergey and Larry are more likable than Bill and Steve. Immigrants, nice-looking guys, a great story. Marissa Mayer: very compelling. Wisconsin, engineer, blonde, future *Vogue* photo feature. It's no accident Google sent Ms. Mayer to Senate hearings to opine regarding the slaughter of newspapers at the hands of Google . . . Oops, I mean the future. When faced with tough questions like "How is the fourth estate going to survive if Google kills the newspaper classified business?" Ms. Mayer responded: "It's still early."[8] Early? It was the two-minute warning in the fourth quarter for newspapers. The gray-haired senators ate it up.

Who wants to be the insurance salesman elected to Congress (the most prevalent career of the House of Representatives)[9] who raises his hand and says, "I'm the guy who does NOT get it. I don't like Apple." Apple is the largest tax avoider in the history of U.S. business,[10] but Apple is hip, and everyone wants to be friends with the cool kid. Same with Amazon, because e-commerce is hip and cool vs. lame and old, traditional retail. In March 2017 Amazon decided to pay sales tax in every state.[11] Here's a company now worth more than Walmart that until 2014 was only paying state sales tax in five states. The benefit of the subsidy has eclipsed $1 billion. Did Amazon need a $1 billion government subsidy? By purposefully managing their business at breakeven, Amazon has built a firm approaching half a trillion dollars in value, that has paid little corporate income tax.

Facebook: Nobody wants to be seen as a company not on board with Facebook. Old CEOs want to put Mark Zuckerberg on stage with his hoodie. It doesn't matter that he is neither charming nor a good speaker—he's the equivalent of skinny jeans and makes every company that tries on Facebook look younger. Sheryl Sandberg also has been key—she's hugely likable, and is seen as the archetype of the modern, successful woman: "Hey everybody! Lean in!"

Facebook has not come under the same scrutiny as Microsoft because it's more likable. Most recently, Facebook has attempted to skirt responsibility for fake news, claiming it's "not a media company, but a platform." Hiding behind freedom of speech and a word, Facebook may have committed involuntary manslaughter of the truth on an unprecedented scale.

It's good to be prom queen.

5. Vertical Integration

The fifth factor in the T Algorithm is the ability to control the consumer experience, at purchase, through vertical integration.

All of the Four control their distribution. If they don't produce the product, they source it, they merchandise it, they retail it, and they support it. Levi's went from $7 billion to $4 billion from 1995 to 2005 because it didn't have control of its distribution. Seeing Levi's jeans piled up as you walk through JCPenney's is just not an aspirational experience. Cartier has caught, or possibly surpassed, Rolex's brand equity by making a big bet on its in-store experience. It turns out that where and how you buy a watch is as important as which tennis player wears that watch. Maybe more.

The ROI of investing in the pre-purchase process (advertising) has declined. That's why successful brands are forward integrating—owning their own stores or shopper marketing. I believe P&G will begin acquiring grocery retail, as they must develop distribution that's growing, and not depend on Amazon, who is their frenemy . . . minus the friend part.

Google controls its point of purchase. In 2000, Google was growing so rapidly that Yahoo, the biggest search engine at the time, bought the rights to offer Google search on Yahoo's homepage. No longer. Facebook obviously is vertical, as is Amazon. Neither produces its own products, but other than sourcing and manufacturing, both control their entire user experience. The biggest innovation for Apple is considered to be the iPhone—but what put the company on track to be a trillion-dollar company was the genius move into retail, usurping control over its distribution and brand. A decision that, at the time, made little or no sense.

A company has to be vertical to reach half a trillion dollars in market valuation. That's easier said than done, and most brands leverage other companies' distribution, as distribution is expensive to build. If you're clothing designer Rebecca Minkoff, you're not going to build your own stores beyond a dozen flagship locations around the world; you don't have that capital. Instead, you sell your products at Macy's and Nordstrom's. Even if you're Nike, it's much more efficient to sell through Foot Locker than to build your own stores.

The Four Horsemen are vertical. Few brands have been able to maintain an aspirational positioning without controlling a large portion of their distribution. Samsung is never going to be that cool, not if it continues to depend on AT&T, Verizon, and Best Buy stores. Remember where you used to go get your Apple computer fixed fifteen years ago? There was a guy who looked like he'd never kissed a girl but was a pro at fantasy adventure games. He stood at a counter in front of piles of gutted computer parts, next to stacks of *Macworld* magazine.

Apple sensed the shift and put people in blue shirts, titled them "genius," and set them in places that brought Apple products to life—spaces whose materials reinforced how special and elegant Apple products are. Apple stores today are intentionally beautiful;

they remind you that Apple, and those who purchase its products, "get it."

6. AI

The sixth factor in the T Algorithm is a company's access to, and facility with, data. A trillion-dollar company must have technology that can learn from human input and register data algorithmically—Himalayas of data that can be fed into algorithms to improve the offering. The technology then uses mathematical optimization that, in a millisecond, not only calibrates the product to customers' personal, immediate needs but improves the product incrementally every time a user is on the platform for other concurrent and future customers.

Marketing historically can be parsed into three major shifts regarding how potential customers were targeted. The first era was *demographic targeting*. Thus, white forty-five-year-old guys living in the city, in theory, will all act, smell, and sound alike, so they must all like the same products. That was the basis of most media buying.

Then, for a hot minute, it went to *social targeting*, in which Facebook tried to convince advertisers if two people, regardless of their demographics, "like" the same brand on Facebook, they are similar and should be grouped/targeted by those advertisers. That turned out to be total bullshit. All it meant was they all shared the action of clicking "Like" on a Facebook brand page, and that was about it—they didn't aspire to the same products and services. Social targeting was a failure.

The new marketing is *behavioral targeting*. And it works:

nothing can predict your future purchases like your current activities. If I'm on the Tiffany website, and I have searched for engagement rings, and I have set up an appointment to purchase such a ring at a certain boutique, that likely means I am about to get married. If I'm spending a ton of time on the Audi site configuring an A4, then I am in the market for a $30,000 to $40,000 four-door luxury sedan.

Thanks to artificial intelligence we now can track behavior at a level and scale previously unimaginable. It's no accident when I start getting served Audi ads all over the web. Behavioral targeting is now the white meat of marketing. The ability to attach behavior to specific identities is the quiet war taking place in media.

There is still a long way to go. I am (as I write this) on a plane from Munich (to Bangkok), where I spoke at the immensely enjoyable DLD conference. DLD is essentially a hip Davos, where followers of the religion of innovation pilgrimage to Munich to worship at the feet of our modern-day apostles—Kalanick, Hastings, Zuckerberg, Schmidt, etc. I can't, obviously, compete with those guys. So, my strategy for getting more attendance and more YouTube views of my talk at DLD? I don a wig and (no joke) dance. I don't play fair—the basis of all good strategy.

In sum, my business strategy message boils down to "What can you do really well that is also really hard?"

First, I say something in my talks. I highlight that Apple is the largest tax avoider in the world because lawmakers treat it like the hot girl on campus—if she pays a little attention to them, they swoon and are willing to enter into an abusive relationship. I say that Uber is fomenting an ethos in business that's terrible for society. Four thousand Uber employees and their investors will split $80 billion (or more) as the 1,600,000 drivers working for Uber will see their wages crash to a level that makes them the working poor. We used to admire firms that created hundreds of thousands of middle- and upper-class jobs; now our heroes are firms that produce a dozen lords and hundreds of thousands of serfs.

The CEOs at events like DLD can't respond to my claims because if they do the markets might listen and the consequences could be dramatic. In addition, they can actually get into serious legal trouble if they, accidentally, disclose nonpublic information. Thus, while I get to put on a show, their speeches are rehearsed and bleached of anything meaningful you haven't read before in a press release from their investor relations department. That's why people attend my talks: I'm free to tell the truth, or at least pursue the truth (I get it wrong all the time).

The CEOs sit and listen to my talks and smile. It's the smile of poker players holding aces. And every one of those aces is data. In the last decade, the world's most important companies have become experts in data—its capture, its analytics, and its use. The power of big data and AI is that it signals the end of sampling and statistics— now you can just track the shopping pattern of *every* customer in *every* one of your stores around the world—and then respond almost instantly with discounts, changes in inventory, store layouts, etc. . . . and do so 24/7/365. Or better yet, you build in the technology to respond every second, automatically. My favorite use of AI is Netflix autoplay for the next episode of a series, which has now been copied by other platforms.

The result is a level of understanding about your customers— indeed, about human nature itself—that has never before been possible. And against smaller, more regional companies it offers a competitive advantage that is essentially unbeatable. The Four have become wizards.

Facility with data, and tech that updates the product real time, will be a key component of the Fifth Horseman. No one has been able to aggregate more intention data on what consumers like than Google. Google not only sees you coming, but sees where you're going. When homicide investigators arrive at a crime scene and there is a suspect—almost always the spouse—they check the suspect's search history for suspicious Google queries (like "how to poison your husband"). I suspect we're going to find that U.S. agencies have been mining Google to understand the intentions of more than some shopper thinking about detergent, but cells looking for fertilizer to build bombs.

Google controls a massive amount of behavioral data. However, the individual identities of users have to be anonymized and, to the best of our knowledge, grouped. People are not comfortable with their name and picture next to a list of all the things they have typed into the Google query box. And for good reasons.

Take a moment to imagine your picture and your name above everything you have typed into that Google search box. You've no doubt typed in some crazy shit that you would rather other people *not* know. So, Google has to aggregate this data, and can only say that people of this age or people of this cohort, on average, type in these sorts of things into their Google search box. Google still has a massive amount of data it can connect, if not to specific identities, to specific groups. And if you don't think they can't find you if they need to, remember: Google also used to claim it erased all of its records on a regular basis. How did that work out?

Facebook can connect specific activities to a lot of specific identities. Facebook has 1 billion daily active users. People live their lives out loud on Facebook, documenting their actions, desires, friends, connections, fears, and purchase intentions. As a result, Facebook is tracking more specific identities than Google, a huge advantage when selling the ability to reach a specific audience.

If I own a hotel in Hong Kong that caters to families, I can go to Facebook and ask for ads targeted to families of a certain income level that travel to Hong Kong at least twice a year. Facebook can identify and serve up the right consumers at a scale previously unimaginable because it can connect data to identity, and we're not as creeped out, as we made this information publicly available ourselves.

Amazon has 350 million credit cards and shopper profiles on file. More than any company on earth, it knows what you like. It's able to connect identity, shopping patterns, and behaviors. Not to be outdone, Apple has a billion credit cards on file and knows the media you most enjoy and, if Apple Pay works, even more than that. Apple too is able to attach purchase data to identity. Owning such a proprietary data set is the Chilean Gold Mines or the Saudi Oil Reserves of the information age.

Just as important, these firms have the skills to leverage software and AI to uncover patterns and improve their offerings. Amazon does an immense number of A/B test emails to see what works best, while Google knows, before anybody else, what you are intending to do. Facebook likely will know more about the arc, texture, and intersection of actions and relationships than any entity in history. What is the endgame, the payoff of this, the greatest aggregation of human talent and data assembled in history? To sell more Keurig Fortissio Lungo Pods.

7. Accelerant

The seventh factor in the T Algorithm is a company's ability to at-tract top talent. This requires being perceived by likely job candi-dates as a *career accelerant.*

The war for tech-enabled talent has reached a fever pitch. A horseman's ability to attract and retain the best employees is the number one issue for all four firms. Their ability to manage their rep-utations, not only among young consumers, but also among their potential workforce, is critical to success. Indeed, one could argue that their brand equity among current and potential employees is *more* important than their consumer equity. Why? The team with the best players attracts cheap capital, innovates, and can spark the upward spiral that pulls away from the competition.

If you're the valedictorian of your class, you have a jet pack strapped to your back in the form of intellect, grit, and emotional intelligence. But you are directionless. You are like Ironman before he learned how to fly—all over the place. A lot of momentum, a lot of thrust, but not much progress. You need to find the right platform to point you in the right direction and accelerate your career.

The Four Horsemen have reputations for doing just that. There are few places a talented twenty-five-year-old can go further by age thirty in terms of role, money, prestige, and opportunities than at one of the Four. The competition to work at one of these companies is brutal. At the nation's military service academies, during one of the first evening meals, it's a common practice to ask the cadets to stand if they accomplished something major in their childhood. Valedictorian? Varsity athlete? Eagle Scout? National Merit Scholar? And as the cadets stand and look around, they are astonished to discover that *everyone* has accomplished those things. The same is true for the Four Horsemen: formidable accomplishments are the *baseline* for applicants. Google is notorious for its vetting process for job applicants, including bizarre questions that have no real answers. The process is the message: if you survive, you are among the elite, the most brilliant members of your generation.

There is no evidence that this process actually works, but that doesn't matter. Getting a job at one of the Four Horsemen is a ticket to the tech illuminati—and the trajectory of your career is about to go vertical.

8. Geography

Geography matters. There are few, if any, firms that have added tens of billions of dollars in the last decade that aren't a bike ride from a world-class technical or engineering teaching university. RIM and Nokia were the pride of their countries, and near the best engineering schools in those countries. The ability to develop and lubricate a pipeline with the best engineering talent from one of the best schools in the world is the eighth factor in the T Algorithm.

Three of the Four Horsemen—Apple, Facebook, and Google— have outstanding relationships with, and are a bike ride away from, a world-class engineering university, Stanford, and short drive to another, UC Berkeley (ranked #2 and #3, respectively).[12] Many would argue the University of Washington (Amazon) is in the same weight class (#23).

To be an accelerant you must have the raw material. Just as you used to build the electricity plant near the coal mine, the raw material today is top engineering, business, and liberal arts graduates.

Tech—software—is eating the world. You need builders, people who can program software, and who have a sense for the intersection of tech and something that adds value to the enterprise and/or the consumer. The best engineers and managers for that task come from, in greater proportions, the best universities.

In addition, two-thirds of the world's GDP growth over the next fifty years will occur in cities. Cities will not only attract the best talent, but manufacture the best talent. The competition and opportunities are similar to rallying with Chris Evert—your game just gets better. In many countries, like the UK and France, one city is responsible for 50 percent of the nation's GDP. Seventy-five percent of large firms are located in what could be called a global supercity. Over the next twenty years, this tendency will likely increase, as firms now need to follow talented young people, not vice versa. Icons of yesteryear are opening urban campuses, prioritizing kids with beards, tattoos, and engineering degrees over people with kids.

It's fairly easy to apply the T Algorithm. I told Nike that to have a shot at a trillion, they would need to do three things:

- Increase percentage of direct-to-consumer retail to 40 percent within ten years (closer to 10 percent in 2016).
- Gain greater facility with data and how to incorporate into product features.
- Move their headquarters from Portland.

As I learned, the algorithm is the easy part. Getting them to listen to you ("You need to relocate HQ from Portland") is the hard part.

Chapter 9

The Fifth Horseman?

LET'S NOW APPLY OUR CHECKLIST of horseman traits to a number of emerging companies that have the potential to become the fifth tech giant. Where are these companies excelling, and where do they fall short? And what will it take to be the Fifth Horseman?

This list of companies is not intended to be comprehensive—after all, great companies regularly appear seemingly out of nowhere thanks to a technological advance, a shift in markets, or a change in demographics—but rather to be broad and thought provoking.

For all they have in common, the Four Horsemen occupy distinct roles in the digital age and have come to prominence through different paths. Two of them, Facebook and Google, dominate categories that did not exist twenty-five years ago. The other two, Amazon and Apple, are in well-established sectors. But while Amazon has overwhelmed its competition through brutally efficient operational prowess and access to cheap capital, Apple led product innovation and secured leadership at the high end—creating entirely new multibillion-dollar product categories and one of the world's great aspirational brands. Facebook had a billion users before its founder turned thirty-two, while Apple took a generation to mature into the globally dominant company it is today.

We should not presume, then, that the next company to emerge as a shaper of the digital age—a Fifth Horseman—will necessarily come from an obviously digital-age industry, or be a highly touted unicorn with a college dropout at the helm. Nor can we presume the next horseman will arise in the United States—although it will certainly have to conquer the U.S. market on its road to success.

We also can't presume that the current Four Horsemen are all guaranteed to hold their positions for decades to come. After all, IBM ruled the electronics world through both the 1950s and 1960s . . . only to lose ground in hardware and, in an amazing feat of leadership, shift to a consulting company. Hewlett-Packard was the biggest tech company in the world just a decade ago . . . only to lose ground under weak leadership, and then be broken up. Microsoft terrified the entire business world, especially tech, and seemed unstoppable in the 1990s. Like the others, it remains a giant company, but no one still thinks of it as an unstoppable juggernaut destined to rule the world.

Still, the current Four Horsemen, as I've tried to explain in the previous chapters, have certain advantages—in products, markets, stock valuation, recruiting, and management (who have assiduously studied why those earlier giants stumbled). That makes it unlikely they will lose their current dominance for a (human) generation or more (famous last words). All have fought to get where they are—and they won't give up their leadership easily. Even when they collide against each other, they seem to make room before the competition gets too extreme. They, for now, seem (somewhat) content to coexist rather than fight to the death.

Now, let's look at the contenders.

Alibaba

In April 2016, a native online commerce company surpassed Walmart to become the world's largest retailer. It was inevitable, but the surprise was that it was not Amazon that bested the Bentonville behemoth—it was Jack Ma's Chinese powerhouse, Alibaba.[1] To be fair, that's in part a function of Alibaba's business model, whereby it

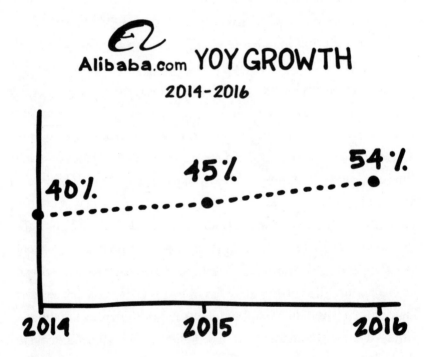

Alibaba Group, FY16-Q3 for the Period Ending December 31, 2016 (filed January 24, 2017), p. 2, from Alibaba Group website.

acts as a marketplace for other retailers—e-commerce and shop-ping, online auctions, money transfers, cloud data services, and a host of other businesses—and it is the $485 billion in "gross mer-chandise value" (GMV) of products sold through Alibaba that beats Walmart. Alibaba itself collects only a fraction of that in revenue—$15 billion in fiscal year 2016.

But size matters, and nobody manages more retail trade than Alibaba. It makes up 63 percent of all China retail commerce, and 54 percent of packages that travel via Chinese post originate from an Alibaba business.[2,3] Alibaba also boasts close to half a billion ac-tive users (443 million) with more monthly active users (MAUs) accessing Alibaba via mobile phones (493 million).[4] Similar to the horsemen, the company has reshaped the retail landscape in China, turning an obscure tradition known as "Singles' Day" (November 11, or "11/11") into the world's largest shopping day. The company did $17.4 billion in GMV on Singles' Day alone in 2016, of which 82 percent originated from mobile.[5]

Alibaba has succeeded because it hit most of the markers we've outlined. It began in a vast market—China—filled with millions of small manufacturers desperate to reach the outside world. It went global almost immediately, reaching almost every country on the planet. It is a master of big data/AI—one of its services. And the market has given it a stratospheric valuation, so it has investment capital to burn. Alibaba has grown so fast, that it essentially has no competition in its corner of the world—as with Amazon, it's easier just to work with Alibaba than to fight it. Many Western brands in China have shuttered their direct-to-consumer sites (unthinkable in the United States and Europe) to focus on their presence on Alibaba and sister property Tmall.

Investors have taken note. In 2014, the company offered what remains the largest IPO in U.S. history, raising $25 billion on a valuation of $200 billion.[6] The stock has underperformed the market since then; however, as I write this in early 2017, BABA has declined 15 percent in value from its offering while Amazon has increased more than 100 percent over the same period.[7]

For all its vast scale, Alibaba faces significant challenges if it wishes to emerge as a global digital-age player in the same class as the Four Horsemen. By definition, it has to expand in a more substantial way beyond its home market—and most important, it has to establish a material commercial presence in the United States, where it operates almost exclusively as an investor. The Chinese market—which seems to grow more volatile by the year—remains as much as 80 percent of Alibaba's business.[8]

As such, Alibaba carries a lot of water on its path to global domination. First, there is no historical precedent for a consumer brand emerging from China. The world is used to global brands from the United States and Europe, and more recently from Japan and South Korea, but not from China. Chinese firms face associations (legitimate or not) of labor exploitation, counterfeit goods, patent infringement, and governmental interference. Those characteristics are inconsistent with the Western values that underpin aspirational brands. And it hasn't helped that Alibaba's early reputation was tainted by claims that many of its small retailers were disreputable.

Ultimately, Alibaba may benefit from the success Apple has had with overcoming concerns about Chinese manufacturing quality, and from other Chinese firms, such as WeChat, developing global followings. Yet the ultimate brand power—an aspirational brand that connotes leadership, luxury quality, and sex appeal—remains a reach for Alibaba. In 2016's list of the hundred most valuable brands, *Forbes* did not include Alibaba.[9]

Alibaba comes up short on visionary capital and has struggled to master storytelling—not just with consumers, but with investors, as Alibaba's opaque governance clouds the story. By comparison, the Four Horsemen are all acknowledged masters at telling their stories, selling their vision, and convincing shareholders to join their Great Crusades. Alibaba, as a conglomerate, doesn't have a real story to tell other than one of continuous success. As we've learned, that's not enough.

Finally, a critical limitation to Alibaba's long-term success is the company's entanglements with the Chinese government. The government has supported its investment in a variety of ways, perhaps most substantially by severely curtailing the operations of Alibaba's U.S. competitors in China.[10] Western investors are willing to accept some level of government interference, but they don't like what seems to be cheating, and the market distortions that result.

While this relationship has doubtlessly been an asset for Alibaba during its growth, investors must be concerned about whose interests will win out when those of global shareholders do not align with the company's superpower patron. Indeed, because of Chinese restrictions on foreign ownership in Chinese assets, foreign investors do not actually own shares in Alibaba, but in a shell company with a contractual right to Alibaba profits—contractual rights enforceable only in Chinese courts.[11] And worse, there are signs that Alibaba cannot count on the support of the Chinese government, with critical stories about the company appearing in Chinese media and from government agencies since 2015.[12,13]

As for the accelerant factor, no doubt working for Alibaba

carries considerable value in China and in other parts of the developing world. But in the West? Not so much. Indeed, it may even prove a stigma—which means that as it moves into the Western markets, Alibaba may find recruiting great talent difficult, and its intellectual capital substandard.

Alibaba's relationship with the Chinese government carries with it the risk that any number of foreign actors, including U.S. and European governments, might see Alibaba through a geopolitical lens and register their concern in the form of regulatory hurdles, investigations, and other roadblocks. These need not be political to be problematic—Jack Ma recently acknowledged that the SEC was investigating Alibaba for various reporting matters related to the firm's complex, multicompany structure. Ma said that "Alibaba's business model does not have any references in the U.S., so it's not just a matter of one or two days for the U.S. to understand Alibaba's business model."[14] That's not exactly heartening.

Finally, data privacy concerns are likely to be a constant thorn for Alibaba as it goes global, limiting its ability to leverage another T Algorithm element, AI.

In sum, the parent brand "China" provides an unwelcome halo of "We may not be cool, but we are corrupt." In high school, the "Bad Boy" who was also lame did not get laid.

Tesla

History is littered with the skeletons of entrepreneurs who challenged big auto—they make movies about them (think *Tucker*). But right now, it looks as if the movie about Elon Musk involves a dope outfit and a brooding Gwyneth Paltrow.

Tesla faces challenges, but it has accomplished more than any other start-up automobile company in our lifetime, and looks well positioned to solidify its position as the market leader in electric-powered cars. Although it remains mostly a luxury product for Silicon Valley bros, its combination of design (no more Hobbit electric cars), innovation in digital control, and massive investment in infrastructure (notably the giant battery factory outside Reno)—not to mention its Edison-like, visionary leader—suggest Tesla has the potential to bust out of its specialty niche and become a mass market player.

Tesla's first volume production car, the Model S, swept the industry awards, garnering the first-ever unanimous selection as *Motor Trend*'s Car of the Year, the highest scoring car *Consumer Reports* ever tested, the "Car of the Century" by *Car and Driver*, and the "Most Important Car Ever" by *Top Gear*.[15] In 2015, it was the highest-selling plug-in electric car in the United States—despite selling for twice the price of its competitors.[16]

The car that has the potential to turn Tesla into an automobile powerhouse is the forthcoming Model 3. Starting at $35,000, it registered 325,000 reservations (requiring a refundable $1,000 deposit) within a week of its announcement.[17] Few firms get access to $325 million in capital for a year at zero borrowing cost. This is a horseman-grade achievement around storytelling.

Still, quite a few variables stand between Tesla today and becoming a Fifth Horseman at some point in the future. Indeed, the company faces challenges beyond those faced by traditional auto companies, as it needs to set up vast networks of charging stations and service stations (where backlogs are an issue), set up global distribution, deal with an array of government subsidies and expectations for electric cars, and manage regulators in the back pocket of the auto industry. However, what appear (now) to be obstacles could end up being the type of analog moats that sustain a giant. Tesla does as well as any current company against the T Algorithm.

Compare Tesla to our criteria. Its product is unparalleled in quality and technical innovation. Tesla is not just an electric car; it's a better car across several dimensions, including a massive and beloved touchscreen-based dashboard, over-the-air software updates (big data/AI), an industry-leading autopilot mode, and design touches (like rethought door handles) that customers love.

Tesla controls the customer experience in a way that no other car company has done, or will be able to do without radical and costly changes. Automobile firms fail the vertical test, as they have pursued a capital-light strategy with independently owned dealerships that are time machines—visiting one is a trip to 1985. These entrenched third-party dealer networks, the limited ability to modify or enhance the vehicle after it has left the factory, and an industry focus on moving the steel off the lot have created a gulf between car companies and consumers.

Tesla's most revolutionary change to the auto industry is not its electric engine—everyone is building those—but its proximity to the customer. From Musk's livestreamed product announcements, to their owned dealerships, to the regular, over-the-air product updates, Tesla understands that a $50,000–$100,000 purchase is the start of a multiyear relationship with Tesla, not John Elway's Claremont Chrysler Dodge Jeep Ram. If Tesla can maintain quality customer support in the face of its rapid growth, Tesla's superior repeat customer rates will become a static part of the story that enables access to cheap capital, which will provide resources to enhance the customer experience, increasing repeat purchases, and so on, and so on.

Tesla trades now at nine times revenue vs. Ford and GM at less than .5 times. In April 2017 Tesla surpassed Ford in market value despite selling 80,000 cars in 2016 vs. Ford's 6.7 million. Tesla has returned to the public market for secondary offerings regularly since its 2010 IPO, most recently raising $1.5 billion to fuel production of the Model 3—despite never recording a profitable quarter.[18] It does so because investors respond to Musk's vision; they buy into his story. This is a guy who says he's going to put rockets into space, revolutionize the car industry, and transform the power storage industry. Oh, and build hypersonic trains on evenings and weekends. What if you could go back and invest in Thomas Edison's ideas? Well, here's your chance.

PRICE:SALES RATIO
APRIL 28, 2017

O
0.29x

O
0.32x

O
6.5x

TESLA

Yahoo! Finance. https://finance.yahoo.com/

Tesla owners describe their purchase decisions in messianic terms and value the company's "mission" above the particulars of its product.[19] But this isn't your hippy uncle's green brand. Tesla is also a luxury brand, and that combination is potent. Every other electric car looks like a Birkenstock; the Tesla looks like a Maserati. No other brand can simultaneously tell people: "You can afford a $100,000 car, you have great taste, *and* you care about the environment." Or put another way, I'm awesome and you should definitely have sex with me. That means, even more than Apple, Tesla has the ability to hit the customer—gently—right smack in the groin.

Don't bet on Tesla limiting itself to automobiles. It already is developing deep expertise in the capture, storage, and transport of electricity. It is putting self-driving auto technology on the road by the tens of thousands while Google and Apple are still in the research park. These are technologies and skills that go beyond personal automobiles and hold the potential for early market leadership in other transport markets, in alternative power generation, and in other uses of electricity in the digital age.

Still, there remain two big obstacles to Tesla in its race to the stable. First, it's not yet a global firm—the majority of its business is done in the United States. Second, Tesla doesn't have a ton of customers, so it doesn't possess data on individual behavior at scale yet. But its cars are data-collecting machines, so the challenge here is scale and execution, not the underlying capability.

Uber

As I write this, around 2 million people drive for Uber (called "Driver Partners"), which is more than the total number of employees of

Delta, United, FedEx, and UPS[20] *combined*. Uber adds 50,000 or more drivers per month.[21] The service is available in more than 81 countries and 581 cities.[22] And it's winning in (most of) those markets.

In Los Angeles, only 30 percent of ride-hailing trips were in taxis in 2016.[23] In New York, almost the same number of cabs and Ubers are hailed daily (327,000 vs. 249,000).[24] For many urban dwellers around the world, Uber has become their default transportation solution, the dominant brand in a space that was previously a hodgepodge of local operators and a penchant for yellow.

These days Uber is the first and last thing I spend money on in every city I visit. Imagine paying $100 every time you entered or left a city or country. That's the relationship the global business person, a very attractive segment, has with Uber . . . or Uber has with us.

I get off the plane in Cannes, France, where I'm speaking at the Cannes Creativity Festival ("The What-Advertising-Sucks-Least Festival"). There's the Uber app on my phone. I see UberX, UberBLACK, and something called UberCopter. My finger dives to that UberCopter button on my phone reflexively—who wouldn't want to know what *this* is? I get a call ten seconds later saying, "Meet me at baggage claim."

They put me in a Mercedes van, drive me half a kilometer to a helipad. I get in a lawnmower with a propeller, piloted by a guy who looks like my paper boy in a pilot's Halloween costume . . . and for 120 euros (about 20 euros more than a cab), I'm choppered over the Côte d'Azur and set down three hundred meters from my hotel. For a moment, I'm James bond . . . minus the looks, skills, gadgets, sex appeal, Aston Martin, and license to kill. Still, close . . .

This is not only supercool but possible, because Uber has access to visionary capital and has paired it with creativity and a lack of respect for the norms around customer experience. The company can do crazy shit like that—decide to take everybody on a helicopter from an airport to a luxury hotel, or deliver kittens on Valentine's Day. But it fails on vertical, as the cars are owned by the drivers, who often work with competitors. Not owning cars has helped them scale fast, but it makes them vulnerable, as they have no analog moats. As you might imagine, Uber also has considerable big data skills—it knows where you are, where you're going, where you're likely to go, and it's all linked to your identity. The app is already auto-populating your destination based on travel history, aging in reverse.

Uber isn't known as much of an accelerant, because very few people know anybody who works for Uber HQ. Uber only has a few thousand employees, and they're very technically literate. Uber has figured out a way to isolate the lords (8,000 employees) from the serfs (2 million drivers), who average $7.75/hour, so its 4,000 employees can carve up $70 billion vs. $2 million on an hourly wage.[25] So, Uber has said to the global workforce, in hushed but clear tones: "Thanks, and fuck you."

Can a car service really justify Uber's $70 billion private-market valuation? Doubtful. But Uber is more than just a car service. In fact, taxis are to Uber what books were to Amazon. It's a real business, and one Uber can do quite well with, but it's only the camel's nose under the tent. The real prize is leveraging its massive driver network (and soon, its massive self-driving car network). In California, Uber trialed UberFRESH, a food delivery service. In Manhattan, it trialed UberRUSH, a package courier. In Washington, D.C., it started UberEssentials, an online ordering and delivery service of grocery store essentials.[26] The firm appears to be building a vascular (last-mile) system for global business—that is, taking the "blood" of commerce to the "organs" of business, globally.

Getting atoms (stuff) around is still a huge issue for firms and people, and Uber could be the equivalent of the transporter from *Star Trek*, only safer and cheaper (if a bit slower). It's likely that, without yet recognizing it, we are seeing a celebrity death match take shape between Uber and Amazon for control of the last mile. Meanwhile, FedEx, UPS, and DHL are about to get a lesson in disruption.

Uber checks almost every box in the T Algorithm: differentiated product, access to visionary capital, global reach, big data skills. That said, beyond execution (no small thing) Uber has only one obstacle, but it is a significant one, to getting to a trillion-dollar valuation: likability. Uber faces challenges on this factor along two fronts.

First, its CEO is an asshole, or at least he's perceived as an asshole. This fact gave rise to a few instances where consumers were encouraged to delete the app, and many did. Where the firm likely lost $10 billion plus in value in forty-eight hours was not the number of people who deleted the app, but the discovery of substitutes, as Uber isn't vertical, and Lyft was able to access many of the same drivers. It's not just the CEO throwing up on himself. In 2014, an Uber senior vice president suggested—in the presence of a journalist—that Uber hire opposition researchers to dig up dirt on journalists who wrote unflattering stories about the company. There have been a series of reports that Uber management uses the technology's ability to track riders in real time for entertainment or other personal reasons, including members of the press.[27] In France,

Uber ran an ad campaign that, at best, was sexist, and arguably suggested that Uber was a great way to hire an escort service.[28] In 2016, Uber paid a $20,000 fine as part of an investigation by the New York attorney general into the misuse of its tracking capability.[29]

Worst of all, Uber's likability took a major hit with Susan Fowler's corporate sexual discrimination charges in February 2017.[30] Actions by midlevel and C-level management ranged from callous to reprehensible in dozens of instances. Scrappy start-ups can sometimes get away with this sort of thing, but industry giants are expected to display greater maturity. Heads should have rolled, and some did, if months later. In June 2017, despite recommendations by external counsel that sought to reallocate the responsibilities of Kalanick, the board initially didn't fire Kalanick; instead he announced he was taking an unlimited leave of absence. The leave of absence narrative showed poor judgment on the part of the board, letting a bad situation get worse. Under pressure from investors, Kalanick resigned the following week. He is clearly a gifted visionary who's built something world changing. But as the firm enters a new stage, it needs a CEO with a new focus and crisis-proof management skills. Uber is now worth more than Volkswagen, Porsche, and Audi, and thousands of families and investors are reliant on the firm and its leadership. This is no longer about Travis, and the firm shouldn't have to see if his frat-rock rehab takes effect or he relapses.

Will this controversy hurt Uber? Yes, but there will be a lag, and not where you think. Consumers talk a big game about social responsibility and then buy phones and little black dresses manufactured in factories where people kill themselves and pour mercury into the water. Uber has an outstanding product, and revenue growth will continue to accelerate. Where it hurts is in the distraction among management, costing them the ability to attract and retain the best talent—where the war is won or lost in a digital age.

Beyond the PR and management crises, Uber's likability risk comes from a more fundamental place than management's bro behavior. Uber is undoubtedly a disruptor in the great tradition of Silicon Valley disruptors. Unfortunately for Uber, the market it's disrupting is a heavily regulated one, and Uber benefits greatly by its attitude that it is not subject to the same regulations as traditional taxis. It believes, and the market has rewarded this belief, that it can hire whomever it wants to drive, and it can charge whatever it wants. Meanwhile, its taxi competition has no such freedom in most markets. Nor does Uber necessarily play fair with its ride-sharing competitors, such as Lyft. There have been several reported incidents of Uber employees engaged in organized efforts to sabotage the competition by ordering and canceling rides from those competitors repeatedly—something like a real-world denial-of-service attack.[31]

At an even broader level, Uber's business model has been attacked for undermining employment relationships and creating unstable, low-wage work that can dry up without recourse. The company maintains that it doesn't run a car service at all, but rather it provides an app that allows drivers to share their cars for a fee.[32] This has raised a host of concerns about driver insurance and benefits, what sort of safety and security obligations Uber has, and other issues.

Hence the #DeleteUber movement that sprung up in minutes in February 2017 and led to an estimated 200,000 Uber users quitting their account with the company over claims that Uber was trying to exploit users during a taxi strike at JFK Airport over protests of President Trump's immigration executive order. The claim was that Uber used the strike to market itself to desperate protesters stuck at the airport. That the story wasn't really true didn't matter—it was a glimpse into the disquiet even loyal users feel about Uber's methods.[33]

The world is still trying to figure out if Uber is good for us, or not. Uber may be a glimpse into what the future looks like in a digital economy: incredible apps providing a remarkable consumer experience subsidized by swooning investors—but also millions of low-paying jobs and a small segment of society splitting a herculean windfall. Thousands of lords, millions of serfs.

Walmart

Walmart may have let Amazon leap to an early lead in the race to be the dominant retailer of the digital age, but it's not out of the race yet. With nearly 12,000 stores in 28 countries, it generated more revenue than any other company in the world in 2015, as it has every year in this century.[34]

When the world was moving online, Walmart was starting to look like a dinosaur. But as companies are realizing that online commerce can only thrive long term when it's embedded in a real-world infrastructure that includes stores, Walmart is still a force to be reckoned with. It has decades of experience managing tight inventories and efficient delivery systems, and its 12,000 stores can be 12,000 warehouses, 12,000 customer service centers, and 12,000 showrooms. Add in that some customers actually *live* in their RVs in Walmart parking lots, and you have a very interesting market advantage.[35]

In late 2016 Walmart acquired Jet.com for $3 billion, or $6.5 million per employee. Jet.com had no viable business model (needed to get to $20 billion in revenue to break even) and was spending $5 million a week on advertising when the deal went through. However, it has a horseman skill: storytelling. Dynamic pricing, as told by the founder of a firm acquired by Amazon, Quidsi, made Marc Lore a potential savior. I believe Jet.com was the equivalent of $3 billion hair plugs purchased by a retailer in a full-blown midlife crisis. However, to be fair, the firm does seem to have gotten its groove back regarding e-commerce. Lore has pushed for operational efficiency, price transparency, and savings via in-store pickup.[36] We'll see.

But Walmart seeking Botox is just the beginning. The firm has access to immense capital, but it's not cheap, as the firm trades at a multiple of profits, which is customary for a retail firm. When the Arkansan retailer announced earnings would take a hit, as they were (rightfully) increasing CapEx to compete with Amazon, the next day the firm shed the equivalent of Macy's from its market cap.

In addition, Walmart is not very likable, as they are the largest employer in the world with more minimum-wage workers than any other U.S. company but also populate the wealthiest people in the world list with a host of Walton kids, who are worth more than the bottom 40 percent of American households. Finally, if you ever wondered who are the people and households that don't own a smartphone or have broadband, look no further. It's the Walmart shopper. The term *late adopter* defines Walmart shoppers. Digital programming and innovation gets less traction with this cohort.

Microsoft

Microsoft is no longer the Beast of Redmond, the company that utterly dominated the PC era. But Windows still powers 90 percent of the installed basis of desktop computers (even if half of those are still creaking along with Windows 7).[37] Office remains the world's default productivity suite, and professional products such as SQL Server and Visual Studio are ubiquitous. If it hadn't so badly failed with its Windows Phone, Microsoft would most likely already be the Fifth Horseman and perhaps still the most powerful company on earth. If it can manage to grow LinkedIn without crushing it in its embrace, Microsoft may still have a chance.

In addition, it has found elusive growth with its cloud offering, Azure. This, coupled with a youthful new CEO, has breathed new life into the Microsoft story. It is no longer the accelerant it used to be, but its focus on the enterprise (vs. consumer for the Four) gives it a marketplace that hasn't seen the same level of innovation or competition as consumer tech.

And its other (growth) story? LinkedIn.

The professional counterpart to Facebook, LinkedIn has some important and tangible advantages compared with its big social counterpart. Facebook gets the bulk of its revenues from one source: advertising. By comparison, LinkedIn has *three* distinct sources of revenues: it sells advertising on its site; charges recruiters for upgraded access to candidates; and sells users premium subscriptions with benefits for job hunting and business development. That's balance. These subscription revenue sources make LinkedIn unique not only with respect to Facebook, but every other major social media player.

LinkedIn also faces an enviable competitive landscape—it has

LINKEDIN REVENUE SOURCES
2015

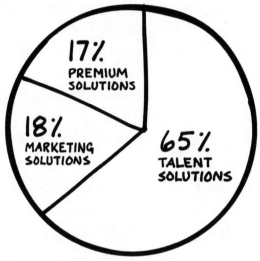

LinkedIn Corporate Communications Team. "LinkedIn Announces Fourth Quarter and
Full Year 2015 Results." LinkedIn.

no true competitor. There are niche sites for specific professions, and Facebook itself represents potential competition, but *nobody* is offering anything like LinkedIn's broad coverage of employment and business networking. You may trade off Facebook for Instagram, Instagram for WeChat, WeChat for Twitter, etc. However, in the B2B world, you are posting your CV on one platform, and that's LinkedIn. You get pissed off at LinkedIn or you decide it's not cool, where do you go? Nowhere. LinkedIn stands alone; it's one of one right now, with no obvious new competition on the horizon.

LinkedIn, by the nature of its business, also has an enviable customer base. More than 467 million people are on LinkedIn, and not just any 400 million.[38] Composed of savvy college grads looking to show off their qualifications and business leaders from around the word—one out of every three people have a LinkedIn profile.[39] So, to the question "Who's on LinkedIn?" the answer is "Anyone who matters." There is a small sliver of Baby Boomer CEOs who aren't on LinkedIn because they are worried they'll be harassed by job-seekers, or they are still trying to figure out their Motorola Razr phone. Other than them, the LinkedIn cohort is global and encom-passing. (By the way, the market for advertising in B2B is twice that of B2C, so the addressable market for LinkedIn is greater than for all the B2C social platforms.)

The trade-off facing LinkedIn, however, is that with focus comes limits. LinkedIn is successful because it serves a relatively narrow market with a relatively narrow set of services. Being the world's directory for professionals is a big business, but it can only be the beginning for a company with aspirations of horseman status.

How LinkedIn builds on that platform is now up to Microsoft. The potential for integration with Outlook and Microsoft's other productivity apps is compelling, not to mention Windows and Mi-crosoft's long-struggling efforts in mobile. But those opportunities may also doom any ambitions LinkedIn has to be a dominant force on its own, since its fortunes are now going to be evaluated based on its ability to drive Microsoft's bottom line. And, for twenty years, that has meant maintaining the omnipresence of Windows and Office at the expense of everything else.

So, the biggest challenge for LinkedIn to reach horseman status is that while the firm checks all the boxes, it checks them in pencil, not ink. Its product is good, but not as good as Facebook. It has access to visionary capital, but not at the same low price as Amazon. And now it's owned by a company that is resurgent, but after more than a decade of decline. In sum, LinkedIn is the Bruce Jenner of this analysis: a great athlete who did a lot of things well . . . after all, Bruce won an Olympic gold medal for the decathlon, and was on the box of the Wheaties I ate in elementary school (sorry, Caitlyn, you'll always be Bruce to me). But Jenner was never a gold medalist in any of those individual sports. He was, to use an old phrase, "A Jack (now Jill) of all trades, but master of none."

Airbnb

It would be tempting to say Airbnb is the Uber for hotels, and move to the next candidate. However, there are stark differences that illuminate Airbnb's competitive strength, relative to Uber, and how the T Algorithm can be used to influence strategy and capital allocation.

While they both are global and enjoy access to cheap capital, their product has substantially different variance. NYU Stern Professor of Management Sonia Marciano (clearest blue-flame thinker in strategy today) believes the key to establishing advantage is finding points of differentiation where there is large, real or perceived, variance. If you're a decathlete, the key is to find the event with the greatest variance in performance and own it. Uber is a great product, but I'd challenge you to identify (without knowing which ridesharing platform you booked through) the difference between Uber, Lyft, Curb, and Didi Chuxing.

The category is a 10x improvement over cabs and black cars, but there is an increasing sameness among ride-sharing players. This has likely been the case for a while, but Uber's CEO frat rock (that is, shit for brains) behavior has prompted people to discover on their own that Lyft is the same thing. The Airbnb platform takes on greater importance as an arbiter of trust, as there is greater variance in the product—a houseboat in Marin vs. a townhouse in South Kensington. United Airlines has more differentiation than Uber right now, as they can drag someone off a plane (due to their fuck-up), but if you need to get from San Francisco to Denver (United hubs), you're going to forgive, because that United flight is highly differentiated (only choice).

In addition, Airbnb has another moat regarding product. Specifically, the liquidity of their product. Liquidity translates to having enough suppliers and customers who can be matched to make the service viable. Both have achieved this. However, the liquidity Airbnb has garnered is more impressive and harder to replicate. Uber needs a mess of drivers and people looking for rides to build a business in a city. Uber's cash hoard gives them the ability to ramp up a city, as can other ride-hailing firms with sufficient capital. However, Airbnb needed to achieve a critical mass of supply in one city and demand (awareness) in many others—people visit Amsterdam from all over the world. There is competition for Uber in every major city, as a firm only needs to establish liquidity in one market. Airbnb needed, and reached, scale on a continental and then global level.

Airbnb's and Uber's valuations (at time of this writing) are $25 billion and $70 billion, respectively. However, I believe Airbnb will surpass Uber's value by the end of 2018, and Uber will register the mother of all write-downs as word spreads regarding their lack of product differentiation and regional competitors take an awful income statement ($3 billion in losses on $5 billion in revenues in 2016) and make it worse.

Airbnb is the most likely "sharing" unicorn to become the Fifth

Horseman. Their weakest point is their lack of vertical integration (they don't own any apartments), meaning Airbnb doesn't have the same degree of control over the customer experience as the Four. This warrants a hard look, by Airbnb management, at allocating some of that cheap capital toward greater control of the channel—long-term exclusives with properties and consistent amenities (wireless, docking stations, local concierge in each metro, etc.).

IBM

Before Google, before Microsoft, before some of the readers of this book were even born, there was one company that mattered in tech. Big Blue *was* technology, the de facto standard for Corporate America and, after joining with Intel and Microsoft, the dominant firm of the first quarter century of personal computers.

But IBM isn't on this list for nostalgia purposes. Even as its revenues continue the long, slow decline from their majestic heights (nineteen straight quarters of declining revenues as of Q1 2017), the company still recorded $80 billion in revenue in 2016, and every year the mix shifts from legacy computer hardware toward high-margin services and recurring relationships.[40] IBM's vaunted sales force can still get meetings with every Fortune 500 CTO, and the company is a serious player in the race to get corporate America into the cloud. IBM has a new, more handsome lead character in their story: Watson. The firm is global and (arguably) vertical. However, the movement up the food chain to services puts them in a business that trades at a multiple of EBITDA vs. revenues, limiting access to cheap capital, and they are seen as a safe place to get a job vs. inspiring. The kids at IBM are the ones who got second-round interviews at Google, but didn't get an offer. IBM is no longer seen as the career accelerant it once was.

Verizon/AT&T/Comcast/Time Warner

This book assumes you are online. And who almost certainly owns the line you are on? One of these four companies. Cable and telco lines were one of the great legal monopolies of the twentieth century, and the big four companies to have emerged from decades of mergers are essential players in the digital age.

They face some major obstacles to taking advantage of their position, however. In particular, most people hate them, and they have no clear path to global status, as local telcos are a source of national identity and governments are pesky about other nations listening in on their phone calls and data. That said, everyone hated the railroads too, and canal boat companies, and stage coach firms. As Ernestine the phone operator, played by Lily Tomlin, used to say, "We don't care. We don't have to. We're the Phone Company."

If you own the pipes on which the world's data travels, you are always going to be important, highly profitable, and very big. That doesn't fulfill many of our criteria for being a horseman, but it may be enough to get you close. After that, all it would take would be an outbreak of enlightened management and to be viewed as an accelerant—unlikely, but still possible.

• • • •

Could any of these companies become the Fifth Horseman? And would the Four Horsemen allow it to happen? Surely Amazon is never going to let Walmart take back all of the ground the younger company has captured. And Google, as it pursues autonomous vehicles, is surely aware of Uber and Tesla.

But there is no accounting for the twists and turns of history. In 1970 IBM seemed unstoppable. In 1990 Microsoft made the electronics industry quake in terror. Companies grow older, success breeds complacency, and the departure of top talent in search of new challenges and pre-IPO options on equity is inevitable. And, of course, there is the wild card: right now, in some lab or dorm room, someone is working on a new technology that will turn the digital world upside down—just as the transistor did in 1947, and the integrated circuit in 1958. Elsewhere, at some kitchen table or a booth at Starbucks, a start-up team, led by the next Steve Jobs, is plotting a new enterprise that could streak past the horsemen to become the first 1-T corporation. It's not likely, but it never is. Like hundred-year floods that seem to be happening every ten years, it seems impossible until it isn't.

Chapter 10

The Four and You

THE DOMINANCE OF THE FOUR has an outsized effect on the competitive landscape and the lives of consumers. But what is their impact on the average career path of the educated individual? I'd argue no young person today should be ignorant of the Four and how they've reshaped the economy. They've made it harder for middle-of-the-road companies to succeed or for any consumer-facing tech start-up to compete and survive.

Given that most of us—and statistics support me on this—are average, what can we learn to help us make the jump from good to strong, even great? I close the book with some observations on what a successful career strategy looks like in this brave new world.

Success and the Insecurity Economy

In sum, it's never been a better time to be exceptional, or a worse time to be average.

That's one of the major effects of the disruptive environment created by the rise of the lottery economy, wherein digital technology creates a single market in which one leader can capture the overwhelming majority of gains. A series of discrete ponds, businesses, and geographies are in the midst of the downpour of globalization, making a smaller number of really big lakes. The bad news: there are more predators. The good news: the big fish in the big pond has a phat life. The Four Horsemen demonstrate this on a mega-scale.

There is a marketplace corollary to this phenomenon, where the value of the top-tier products in a category explodes, even as the value of lesser products collapses. In rare books, Amazon has given once-obscure and hard-to-find editions global exposure. Predictably, the resulting increase in demand for a fixed supply has led to higher prices—for the finest masterworks. But it has also illuminated the abundance of run-of-the-mill books and given the buyer exponentially more choices below the top tier. Which, just as predictably, has had the opposite effect, crushing the value of these non-top-tier books.

The same thing is happening in labor markets. Thanks to LinkedIn, everyone is on the global job market all the time. If you are exceptional, there are thousands of firms looking for, and finding, you. If you are good, you are now competing with tens of millions of other "good" candidates all over the planet—and your wages may stagnate or decline.

The top dozen professors at Stern are in demand globally and get paid $50,000 or more to speak at a lunch. I'd venture their average annual income is $1 million to $3 million. The rest ("good") are now competing with Khan Academy and the University of Adelaide (both offer "good," the former online). These "good" professors teach executive education for modest extra income, or complain about the dean in a primal scream for relevance, as they make a fraction of what their (marginally) better colleagues make. The difference between good and great can be 10 percent or less, but the delta in rewards is closer to 10 times. The "good" professor's average annual income is $120,000 to $300,000, and they are overpaid—and easily replaced. The university can't fire them, thanks to tenure, so it pretends to be concerned and (mostly) ignores them. It makes them department chairs, assigns them to committees, and comes up with a host of excuses for their mediocrity.

So, if not naturally great, what behaviors help achieve the extra 10 percent? The fundamentals won't change. Excellence, grit, and empathy are timeless attributes of successful people in every field. But as the pace and variability of work increase, success will be at the margins, separating successful people from the herd.

As I described at the beginning of this book, my sixth company is L2, a *business intelligence* (fancy term for research) firm that has grown to 140 people in seven years. Seventy percent of our employees are under thirty; the average age is twenty-eight. L2 employees are often recruited by aspirational firms. They are kids: raw, having had little time to shape their working personalities beyond the nature and the nurture of their youth. It's an interesting environment to observe people and witness how their core personalities drive success and failure. And from those observations, I've come to some conclusions regarding what it takes to succeed in our evolving, horsemen-driven economy.

Personal Success Factors

On average, smart people who work hard and treat people well do better than people whose thinking is muddled, who are lazy, or who are unpleasant to colleagues. That has always been and will always be true—even if the occasional jerk proves the exception. However, talent and hard work only get you in the top billion on the planet. There are other, more subtle centrifuges and separators that create the cream of the digital age.

Nothing is more important than *emotional maturity*—especially for people in their twenties, in whom this quality can vary widely. There are fewer and fewer fields in which a person reports to work with a single boss, a specific set of tasks, and the expectation that those parameters won't change frequently or significantly. By comparison, the digital-age worker must often respond to numerous stakeholders and shift between roles throughout the day—an environment that favors the mature. And as competitive and product cycles shorten, our work life will see rapid swings between success and failure.

How well someone manages their own enthusiasm through those cycles is important. How people interact with one another determines the projects they work on, who will work with them, and who wants to hire them. Young people who have a strong sense of their own identity, remain poised under stress, and learn and apply what they've learned, do better than peers who are more easily flustered, get hung up on petty issues, and let their emotions drive their responses to stimuli. People who are comfortable taking direction and giving it, and who understand their role in a group, do better than their peers when lines of authority get murky and organizational structures are fluid.

This effect has been well documented in the academic environment. A massive meta-study of 668 evaluative studies of school programs teaching social and emotional life skills found that 50 percent of children in those programs increased their scholastic achievement, and there were similarly dramatic drops in misbehavior. And best-selling author Daniel Goleman, who popularized the term *emotional intelligence*, found measurable business results at global companies led by individuals who demonstrate self-awareness, self-regulation, motivation, empathy, and social skills.

One interesting result of the increasing importance of emotional maturity is that among younger people, this skill favors women. I'm not trying to be politically correct here, though admittedly I'm not sure I would have had the balls to highlight this point if the finding favored men. Anyway, when asked in surveys, men and women agree that women in their twenties tend to "act their age" more than men. There is neurological evidence that women's brains develop sooner and more quickly into adult brains.

I often attend meetings where a young man, or several, burn up most of the time expounding on their own enthusiasms, clash over perceived control of the dialog, and generally preen before the crowd, until finally a young woman in the room—who has kept her mouth shut and listened—calmly introduces relevant facts, summarizes the critical issues, and makes the recommendation that gets us on to our next task.

Men, even young men, still enjoy a cultural bias over their female peers when it comes to advancement—probably because they are seen as more decisive. This will likely remain the case for that minority of young men who cultivate emotional maturity. But they will be a rare and valuable breed. Firms have figured out that, with

70 percent of high school valedictorians female, the future really is women.

The digital age is Heraclitus on steroids: change is a daily constant. In almost every professional environment, we are expected to use and master tools that did not exist a decade ago, or even last year. For better or worse (and frankly, it is often for worse), organizations have access, essentially, to infinite amounts of data, and what might as well be an infinite variety of ways to sort through and act on that data. At the same time, ideas can be turned into reality at unprecedented speed. The thing Amazon, Facebook, and no less hot firms, including Zara, have in common is they are *agile* (the new-economy term for fast).

Curiosity is crucial to success. What worked yesterday is out-of-date today and forgotten tomorrow—replaced by a new tool or technique we haven't yet heard of. Consider that the telephone took 75 years to reach 50 million users, whereas television was in 50 million households within 13 years, the internet in 4, . . . and Angry Birds in 35 days. In the tech era, the pace is accelerating further: it took Microsoft Office 22 years to reach a billion users, but Gmail only 12, and Facebook 9. Trying to resist this tide of change will drown you. Successful people in the digital age are those who go to work every day, not dreading the next change, but asking, "What if we did it this way?" Adherence to process, or how we've always done it, is the Achilles' heel of big firms and sepsis for careers. Be the gal who comes up with practical *and* bat-shit crazy ideas worth discussing and trying. Play offense: for every four things you're asked to do, offer one deliverable or idea that was not asked for.

Another standout skill is *ownership.* Be more obsessed with the details than anybody on your team and what needs to get

THE MARCH TO A BILLION USERS

YEARS

■ Office	1999	2011		21.7
✉	2004	2016		11.8
f	2004	2013		8.7

Desjardins, Jeff. "Timeline: The March to a Billion Users [Chart]." Visual Capitalist.

done, if, when, and how. Assume nothing will happen unless you are all over everybody and everything, as it likely won't. Be an owner, in every sense of the word—your task, your project, your business. You own it.

Go to College

Yeah, I know . . . no shit. Still, it bears repeating. If you want to be a white-collar success in the digital age, the clearest signal is attendance at a prestigious undergraduate school. And the distinction matters.

Yes, Zuckerberg, Gates, and Jobs all dropped out of college. However, you, or your son, are not Mark Zuckerberg. And while none of them graduated, their college experiences were still instrumental in their success. Facebook went viral among college students because it grew out of a real need on campus. Gates spent three years intensely studying math and programming at Harvard before he started

Microsoft, and he met Steve Ballmer there, the man to whom he'd turn over the reins of Microsoft a quarter century later. And even Jobs, who passed through Reed College in an adolescent daze, famously had his passion for design sparked there. All the bullshit, cost, and stress parents endure to get, and keep, their kids on the path to a decent four-year school is still, very much, worth it. College grads make ten times more, over their lifetime, than people with just high school degrees.

There are precious few places in the world and times in our life when we are put in the simultaneous presence of eager and bright young minds, brilliant thinkers, and the luxury of time to mature and generally ponder the opportunities set forth by the universe.

So, go to college—you may even learn something. But even if you don't, a brand-name college on your forehead will be your greatest asset until you have assets, and it will never stop opening doors. HR departments, graduate program admissions committees, and even potential mates are busy people with lots of options. We all need filtering mechanisms and simple rules of thumb to wade through our choices, and it's just too easy to think "Yale = smart; U. of Nowhere = not as smart." And in a digital age, smart is sexy.

No one likes to admit it, but the United States has a caste system: it's called college. At the height of the Great Recession, unemployment among college grads was less than 5 percent, while those with only high school diplomas suffered unemployment rates above 15 percent. And your degree of success is stratified based on the college you attend. The kids who get into the top twenty schools are fine. They can pay off their student debt. Meanwhile, everybody else incurs the same level of student debt, yet faces nowhere near the same opportunities for an ROI on that debt.

The cost of college has skyrocketed in recent years, at a rate of 197 percent vs. the 1.37 percent inflation rate.[1,2] Education is ripe for disruption. There's a commonly believed fallacy right now that technology companies, specifically VC-backed technology education companies, are going to disrupt education. That's bullshit. Instead, Harvard, Yale, MIT, and Stanford are the favorites to disrupt education when they fall under heavy and sustained government pressure over the irrational and immoral hoarding of their mammoth endowments. Harvard claims it *could* have doubled the size of its freshman class last year with no sacrifice to its educational quality. Good. Do it. More students, paying no tuition, at the best schools will disrupt the system, not Massive Open Online Campuses (MOOCs) at mediocre colleges. (See Apple chapter: hope they do it.)

At a top university, the brand isn't the only thing that you'll get besides your education. The friends you make on campus can be just as valuable. Some of those friends will drop off the face of the earth, sure, but some of them will go on to acquire assets, or skills, or connections of their own that, properly networked, may be just what you need to succeed in your own future endeavors. Some of my most trusted advisors and business partners are people I met at UCLA, and later at Haas. I know I would not have had the success I've had without those experiences and friendships.

The problem with this advice, and I'll be the first to admit it, is that it's unfair. The cost of college is ruinously expensive; four years' tuition, plus room and board, at even a second-tier school can run you a quarter million dollars. And though many top-tier schools can offer generous financial-aid packages—financial aid at Ivy League schools, for example, is already so substantial that kids from average income households already get not only free tuition, they get free room and board—it often isn't tuition that keeps bright poor kids out of the best schools. To take advantage of these programs, those bright poor kids have to get admitted, and that means competing with kids who've had private tutors, SAT prep classes, and every field trip imaginable. They also have to compete against "legacies"— kids whose parents are alumni of that school. And they have to compete against kids whose parents have been donating money to the school for years, and who play golf with the dean.

If you can't get into a fancy college, what should you to do? Transfer. In most cases, it's a whole lot easier to get into a good school as a junior, where dropouts have left empty slots, rather than as a freshman, where you're up against everybody. Get into a second- or even third-tier university . . . and then work your ass off: a great GPA, honors programs, awards, service clubs, etc. This is also a much cheaper route as well.

Certification

Needless to say, not everyone should go to college, for one reason or another. So, if college is not an option, what to do? Seek certification. A CFA, CPA, Union Card, Pilot's Instrument Rating, RN, Jivamukti Yoga Teacher Certification . . . hell, a smartphone and a driver's license are credentials that distinguish you. College is the most athletic and agile of certifications. If college is not your jam, you need to find other credentialing to separate you from the other 7 billion people on the planet whose average pay is $1.30/hour.

The Accomplishment Habit

People who achieve goals in one area achieve them in all areas. Whether it's making the finals in Division 3 field hockey, winning your elementary school spelling bee, or having an oak leaf cluster pinned on the shoulder of your army uniform, accomplishment is a habit that can be cultivated and repeated.

Winners, first and foremost, have to be competitors. You cannot win without stepping on the field, and it's only by taking that risk (you may get beaned in the face), exposing yourself to failure, that real accomplishment is achieved. Competing requires bravery and action-orientation. Steve Jobs took a lot of grief when he returned to Apple at the turn of the century and announced that he only hired As, because As only hired As, while Bs hired Cs—but he was right: winners recognize other winners, while also-rans can be threatened by competitors.

Competing takes grit. The nonglamour sports academic competitions (crew, gymnastics, water polo, track) are also a breeding ground of competitive grit—the subject of a lot of attention in business books, incidentally. If you can row 2,000 meters after throwing up at 800 and beginning to lose consciousness at 1,400 meters, then you can manage a difficult client and summon the will to push something from good to great.

Get to a City

For years, we believed the digital age would enable us to "work anywhere"—a utopia of people living in quiet mountain cabins, tapping away at their laptops through the magic of the information superhighway. In fact, the opposite has happened. Wealth, information, power, and opportunities have *concentrated*, as innovation is a function of ideas having sex. Progress is typically in person. Also, we are hunter-gatherers and are happiest and most productive when in the company of others and in motion.[3]

More than 80 percent of the world's GDP is generated in cities, and 72 percent of cities outperform their own countries in growth. Every year, a greater percentage of GDP moves to cities, and it will continue to do so. Thirty-six of the hundred largest economies in the world are U.S. metropolitan areas, and in 2012, 92 percent of jobs created and 89 percent of GDP growth came from those same cities. And not all cities are equal—the global economic capitals are becoming supercities. New York and London consistently rank as some of, if not *the,* most powerful cities in the world. Developers are also keen to invest in wealthier cities, where they can expand accordingly (think of Manhattan businesses that are expanding to Brooklyn locations). It appears that the lottery economy applies to real estate, too.

A decent proxy for a twenty-something's success will be their geographic trajectory. How long did it take them to get to the biggest city in their country, then to the biggest on the continent? The strongest signal of success will likely be those who moved to global economic capitals, the supercities, versus those who stayed in the relative hinterlands.

Pimp Your Career

Okay, so you're emotionally mature, curious, and have grit, but you're not the only one. How do you separate yourself from all the other bright young things? First, you need to push the limits of your comfort zone by consistently pimping your attributes. First question: What's your medium? For beer, it's TV; for luxury brands, it's print. What environment is ideal to express "you"? There's Instagram, YouTube, Twitter, firm sports teams, speeches, books (we'll see), YPO, alcohol (yes, it's a medium if you're good at it—fun/charming), or food.

You need a medium to spread your awesomeness, as the path to under-compensation is doing good work that never gets explicitly pimped or attached to you. Yes, it's unseemly, and your work and achievements should speak for themselves. They don't. Figure out how you are going to reach 10, 1,000, 10,000 people who otherwise wouldn't have been exposed to your work and awesomeness. The good news: social media was built for this. The bad news: it's hand-to-hand combat. I have 58,000 Twitter followers, which is good but not great, and it's taken me six years, fifteen minutes a day, to get there. Our weekly "Winners & Losers" videos now get 400,000-plus views a week. Our first, 138 weeks ago, got 785 views. Btw, it's not me and my nine-year-old in the kitchen with a camcorder. Animators, editors, researchers, a studio, and substantial media (that is, we buy distribution/views) have been constant investments over the last 2.5 years so we could become an overnight success.

Some people are better at words, some at images. Invest aggressively in your strength(s) and spend modest effort to get your weaknesses to average so they don't hold you back. Everyone from employers to coworkers to potential mates is looking you up. Make sure what they see is the best of you. Google yourself, and if your feeds can be cleaner, stronger, and more fun, make improvements.

Boom(er)

What if you're not twenty-five and from an Ivy League—turn on the car and close the garage door? No, not yet. I'm fifty-two, working with a workforce that's on average a quarter century younger. There are a few of us old folks at L2. However, we all have one thing in common. We have learned how to manage young people (clear objectives, metrics, invest in them, empathy) and push our comfort zones with the Four—we make an effort to understand and leverage them. The fifty-five-year-old who says (proudly) he or she doesn't use social media has given up or is just afraid.

Get in the game. Download and use apps. Use every social media platform (okay, not Snapchat, you're too old) and, more important, try to understand them (best practices, user reviews, Instagram vs. Instagram Stories). Buy some keywords and post a video on Google and YouTube. No manager says "I don't like business." The Four are business; nothing is immune, and if you don't get them, you (increasingly) don't get business today.

Despite the airbrushed version of me presented on Wikipedia and my NYU Stern bio, I do not take naturally to technology. However, I am passionate about being relevant and creating economic security for me and my family. So, I am on Facebook and, sort of, understand it. My preference would be to post a banner across my Facebook homepage (is that what they even call it?) that reads "There's a reason we haven't stayed in touch." Instead, I try to understand what a "dark post" is and then ping over to Instagram, click on ads, and try to understand why brands are spending less on TV (which I understand) and more on the visual platform. Using and understanding the Four is table stakes. Get. In. The. Game.

Equity and the Plan

Try to get equity as part of compensation (if you don't think equity in your employer is going to be valuable, find a new employer), and increase that ratio (ideally) to 10 percent and 20 percent plus of your compensation by the time you are thirty and forty, respectively. If there are no equity opportunities at your company, you need to create your own by maxing out all tax-advantaged accounts (401k and others) and charting a path to $1–3–5 million, based on your income and spend levels. Time goes strangely slow, and fast. The part where you wake up at fifty without economic security can happen fast. Assume you won't make huge dollars or buy a stock that goes up a hundred times, and, as early as possible, begin building that egg.

I've had several multimillion exits and still managed to wake up, as I had failed to chart a patch, one September morning in 2008 with almost no money. This was about the same time I started having kids, and it was fucking scary. Avoid the fucking scary, and chart a (Plan B) path . . . early. Except when in school, spend less than you make. The happiest people I know are ones who live beneath their means, as they don't have the constant ringing in their ears of economic anxiety. Note: I recognize that for many or most middle-class families this may just not be feasible.

Nobody becomes superwealthy through paychecks—it takes equity in growing assets to create real wealth. Just compare the net worth of CEOs to the founders of their companies. Cash compensation will improve your lifestyle, but not your wealth—it isn't enough, and saving is counterintuitive and just plain hard. High-income individuals tend to flock together, and what we see, we covet. It's surprisingly easy to get used to business class. The definition of rich is when your passive income exceeds your nut (what you need to live). My dad, collecting $45,000 in social security and cash flow from investments, is rich, as he spends $40,000/year. I have several friends in finance who make seven figures who are not rich, as the moment they stop working, they are shit out of luck. The path to rich(es) is a path of living below your means and investing in income-producing assets. Rich is more a function of discipline than how much you make.

Human beings, especially Americans, aren't natural savers. We're optimists, and worse, we tend to view our greatest earning years as normal, and assume any good run of income will last. There are a disturbing number of service industry professionals, athletes, and entertainers who made millions in just a few years, but who ended up broke because they had no forced savings. *Sports Illustrated* estimated that 78 percent of all NFL players are either under substantial fiscal stress or go bankrupt within two years after their career ends.

Serial Monogamy

Familiarity breeds contempt. External hires are paid nearly 20 percent more than company veterans at the same level, despite receiving lower performance evaluations and still being more likely to quit. There is, of course, a balance. If you spend all day shining up your LinkedIn profile and lunching with headhunters, you'll be seen as promiscuous and won't be attractive to any employer.

The strategy is serial monogamy. Find a good employer where you can learn new skills, garner senior-level sponsorship (somebody who will fight for you), get equity/forced savings, and fully dedicate yourself to that company for three to five years. Don't burn mental energy on your external options unless your current situation is awful. Btw, make sure your definition of awful is shared by trusted mentors after describing the "injustices" you are enduring. You should avoid the appearance of actively looking, but be always open to a conversation.

At a sensible juncture (don't start looking when you've just started a demanding new position at your current employer, for example) return headhunter calls, go on some interviews, ask others for help or introductions. Consider if you would benefit from additional training.

If a conversation turns to an attractive offer, be transparent with your current boss—you've been a loyal employee, you like where you are, but you have an offer that is better on xyz dimensions. You are attractive to strangers, as evidenced by the feedback received from the marketplace. Don't bluff. The truth has a nice ring to it. Often your external offer will make you much more attractive to your current firm without having to leave. If your firm does not counter, that means there was limited upside, and it's time to leave. If, on the other hand, this walk on the wild side turns out, settle on the best thing for your next three to five years, then repeat the process.

Stay Loyal to People, Not Organizations

Mitt Romney was wrong—corporations aren't people. As British Lord Chancellor Edward Thurow observed more than two centuries ago, business enterprises "have neither bodies to be punished, nor souls to be condemned." As such, they do not deserve your affection or your loyalty, nor can they repay it in kind. Churches, countries, and even the occasional private firm have been touting loyalty to abstract organizations for centuries, usually as a ploy to convince young people to do brave and foolish things like go to war so old people can keep their land and treasure. It. Is. Bullshit. The most impressive students in my class are the young men and women who have served their country. We benefit (hugely) from their loyalty to our country, but I don't think we (the United States) pay them their due. I believe it's a bad trade for them.

Be loyal to people. People transcend corporations, and people, unlike corporations, value loyalty. Good leaders know they are only as good as the team standing behind them—and once they have forged a bond of trust with someone, will do whatever it takes to keep that person happy and on their team. If your boss isn't fighting for you, you either have a bad boss or you are a bad employee.

Manage Your Career

Take responsibility for your own career, and manage it. People will tell you to "follow your passion." This, again, is bullshit. I would like to be quarterback for the New York Jets. I'm tall, have a good arm, decent leadership skills, and would enjoy owning car dealerships after my knees go. However, I have marginal athletic ability—learned this fast at UCLA. People who tell you to follow your passion are already rich.

Don't follow your passion, follow your *talent*. Determine what you are good at (early), and commit to becoming great at it. You don't have to love it, just don't hate it. If practice takes you from good to great, the recognition and compensation you will command will make you start to love it. And, ultimately, you will be able to shape your career and your specialty to focus on the aspects you enjoy the most. And if not—make good money and then go follow your passion. No kid dreams of being a tax accountant. However, the best tax accountants on the planet fly first class and marry people better looking than themselves—both things they are likely to be passionate about.

Seeking Justice

If you are seeking justice, you won't find it in the corporate world. You will be treated unfairly and will be in unworkable situations that are not your fault. Expect that a certain amount of failure is out of your control, and recognize you may need to endure it or move on. If you leave, keep in mind people remember more about how you leave than what you did while there. No matter the situation, be gracious.

The best revenge is living better than, or at least never again thinking about, the person who made your life miserable. And ten years later, that person might be in a position to help you, or just not get in the way. People who complain about others and how they got screwed are, well, losers. Note: if you believe someone has treated you unethically (such as harassment), don't be afraid to speak to a lawyer and mentors about what to do (there's no one size fits all here).

Regression to the Mean

Nothing is ever as good or bad as it seems. All situations and emotions pass. When you have a big victory, pull in your horns and be risk avoidant for a period. Regression to the mean is a powerful force, and the good luck (and a lot of it is luck) will cut the other way at some point. So, many entrepreneurs who make a lot of money on one venture turn around and lose a lot of it because they believe the victory was due to their genius and they should go bigger. At the same time, when beaten down, realize you are not as stupid as the world, at that moment, seems to think you are. When beaned in the face, the key is to get up, dust off, and swing harder. I've been hit in the face several times, and kept getting up. Also, a couple times, I was looking at private jets (during economic booms/bubbles), only to have the universe remind me I wasn't that smart. However, I've achieved Mosaic status on JetBlue.

Go Where Your Skill Is Valued

Within your organization, figure out what the company is good at—its core functions—and if you want to excel there, have a bias toward one of those categories. Google is all about engineers: the salesmen don't do as well (though it's still a great place to work). Consumer packaged goods companies are brand managers: engineers rarely make it to the C-suite. If you're in the discipline that drives the company, what it excels at, you will be working with the best people on the most challenging projects, and are more likely to be noticed by senior management. This doesn't mean you can't be successful in a cost center, or that you have to make the thing the company sells. Look at the resumes of the senior executives—if they mostly came from sales, then the company values sales. If they are operational people, that's the heart of the firm, whatever it says in the ads.

Sexy Job vs. ROI

Sectors are asset classes—the cool ones are overinvested, driving down returns on human capital (compensation for working there). If you want to work for *Vogue*, produce films, or open a restaurant, you had better get immense psychological reward from your gig, as the comp, and returns on your efforts, will likely suck. Competition will be fierce, and even if you manage to get in, you'll be easily replaceable, as there are always younger, hipper candidates nipping at your heels. Very few high-school graduates dream of working for Exxon, but a big firm in a large sector would give you a career trajectory with regular promotions a sexy industry won't. Job stability counts if you want to have kids. You don't want to be forty-five and worried about your prospects. Join a band on weekends. Learn photography at night. Work on it a little at a time, until you have a nest egg to unleash it fully. The sooner you start earning good money, the less you'll need to earn, thanks to compound interest. In a sexy industry, rent due can drive you to desperation, and you'll have neither a career, a stable future, nor recognized genius.

I don't invest in smoothie bars, new fashion lines, or music

PROF. GALLOWAY CAREER ADVICE

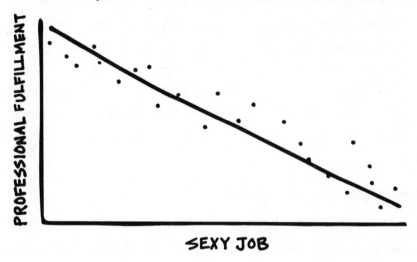

labels. My greatest success has been a research firm. When I have someone smart in front of me excited about a SaaS platform that offers hospitals a better scheduling solution (so boring I want to put a gun in my mouth), I smell money.

Strength

A decent proxy for your success will be your ratio of sweating to watching others sweat (watching sports on TV). It's not about being skinny or ripped, but committing to being strong physically and mentally. The trait most common in CEOs is a regular exercise regime. Walking into any conference room and feeling that, if shit got real, you could kill and eat the others gives you an edge and confidence (note: don't do this).

If you keep physically fit, you'll be less prone to depression, think more clearly, sleep better, and broaden your pool of potential mates. On a regular basis, at work, demonstrate both your physical and mental strength—your grit. Work an eighty-hour week, be the calm one in face of stress, attack a big problem with sheer brute force and energy. People will notice. At Morgan Stanley, the analysts pulled all-nighters weekly, and it didn't kill us, but made us stronger. This approach to work, however, as you get older, can in fact kill you. So, do it early.

Ask for and Give Help

I had a number of uber-successful men in their fifties and sixties who helped me when I was coming of professional age in the nineties in San Francisco (Tully Friedman, Warren Hellman, Hamid Moghadan, Paul Stephens, Bob Swanson). They didn't do it because they knew my parents or thought I was awesome, but because I asked. Most successful people have the time to reflect on important questions, including "Why am I here and what mark do I want to leave?" The answer usually involves helping others. You need to ask for help if you plan on being successful. You also should get in the habit of helping people junior to you. Helping people senior to you isn't helping, but ass-kissing. Expect that a large portion of the people you help will not reciprocate, and you won't be disappointed. However, plant enough seeds helping others, and a few will pay off hugely where you least expect it. It also just feels good.

What Part of the Alphabet Are You?

The different stages of a firm's life cycle require different leadership. Start-up, growth, maturity, and decline require (crudely speaking) an entrepreneur, visionary, operator, and pragmatist, respectively. Surprisingly, the hardest to find are the pragmatists. The entrepreneur is the storyteller/salesperson who convinces people to join or invest in a company before it really exists. At the outset, no company makes sense, or it would already exist. The visionary does the same thing with the company's first, unproven, products or services—even though there is no evidence the company will survive long enough to support those products.

I've started several firms. That makes me, in Silicon Valley's terms, a serial entrepreneur. Serial entrepreneurs share three qualities:

- a higher tolerance for risk
- can sell
- too stupid to know they are going to fail

Rinse and repeat, over and over again.

Highly rational and intelligent people are usually not good entrepreneurs, especially serial entrepreneurs, as they can clearly see the risks.

Once a firm has momentum and access to capital, it is better served by a visionary who can turn this momentum into a somewhat dumbed-down, scalable, and repeatable process and gain access to cheaper and cheaper capital. Entrepreneurs are usually enamored with the preciousness of their product vs. something that can scale. Like the entrepreneur, the visionary needs to sell the story, but it's now a narrative a few chapters in. A visionary may not have the crazy genius of the entrepreneur, but they make up for it with a feel for the organization, specifically the hard work of building an organization that can scale the idea. Once we get to a hundred people, I've always brought in an "organizational" person, as I don't have these skills.

The operator is long on business maturity and reeks of integrity. He or she must be highly competent at dealing with employees who increasingly choose job security over risk, and who prefer salaries over stock. This is the CEO who travels 250 days a year visiting far-flung divisions, deals with angry shareholders, and is always on the hunt for the next corporate acquisition. People who envy high-paid corporate CEOs don't know what they are talking about (other than the tens of millions in comp); it's one of the shittiest jobs in corporate life, which is why certain sociopaths thrive at it.

If the employees and shareholders of an aging and declining firm are lucky, they get a pragmatist in the chief executive's chair. The pragmatist CEO has no romantic notions about the company's glory days (mostly because he or she wasn't there) and *never* falls in love with the firm. Rather, the pragmatist CEO recognizes that the firm is in decline and harvests the cash flows, cuts costs faster than revenue declines, sells off still-valuable assets to mature company CEOs (never to visionary CEOs, who don't want the stink of death on their companies), and then fire-sales the rest.

A productive exercise for one's own career is to ask: Where do I thrive in the alphabet? Think of companies and products having a life cycle, A–Z. Are you happiest at start-ups where you're expected to wear a number of different hats (A–D), the inception/visionary stage (E–H), good at managing, scaling, and reinventing (I–P) . . . or can you manage a firm/product in decline, and do so profitably (Q–Z)?

Few people are good across more than several letters. This exercise should help guide the firms and projects you work for and pursue.

Few CEOs are suited for more than two of the stages. Most CEOs got to the position by being founders, visionaries, or operators, not pragmatists. You can probably count the number of CEOs in American business history who have effectively led their firms (or wanted to) across the entire alphabet. After all, who wants to lead into death the great company they founded decades before?

Kids born today in advanced nations have a life expectancy of one hundred. Of the Dow 100, only eleven are more than one hundred years old—89 percent mortality rate. That means our kids will outlive almost all the firms you know today. Look at the list of the ten largest firms in Silicon Valley for each decade of the last sixty years. It's a rare firm that makes the list twice.

A more likely fate is that of Yahoo—a one-time superstar sold for a fraction of its value a decade ago. Yahoo! (that exclamation point now seems more ironic than descriptive) is stuck in the age of display advertising—and has demonstrated no evidence it is able to do anything else. With a pragmatist running the firm, it could have aged gracefully, reducing the number of employees and divesting noncore assets, producing gobs of cash for loyal investors. When a profitable firm starts reducing expenses versus reinvesting in growth, it can become massively cash generative. Oath is now the property of an old-economy firm, a gray if not a white flag.

Botox

People who received a great deal of attention for their looks at a young age are more likely to opt for cosmetic procedures when older.

It's the same in business. Firms that garnered most of their confidence (valuation) from the fact they were at one time "hot" opt for the equivalent of expensive Botox procedures and eyebrow lifts—acquisition of dubious start-ups (like Yahoo's billion-dollar bet on Tumblr), delusional strategies in mobile computing, hiring expensive talent from younger firms who, like gigolos, take their money and quickly move on—in the doomed hope of recapturing their lost youth. The result is a freakish-looking internet company hopped up on Botox and fillers. Firms in old-economy or niche sectors seem to have an easier time coming to grips with aging and aren't as susceptible to the kind of midlife crises that are expensive and create a great deal of misery for stakeholders.

It's difficult to find pragmatists to run these companies at the end of the alphabet, but they are out there. They can be activist shareholders or partners in private equity firms who have seen firms die and realize that there are worse things than death—specifically a slow death where shareholders are bankrupted trying to give Pop-Pop just one more day. Pragmatists can make unemotional, even cold, decisions to move Nana home and enjoy her last days (that is, return a shitload of cash to investors).

David Carey, CEO of Hearst Magazines, is one of the few CEOs I've seen make the transition from visionary to operator to pragmatist. It's not a shocker that magazines are in structural decline. David hasn't given up hope and regularly launches (surprisingly successful) new titles and has developed profitable digital channels. However, this is pushing a rock up a hill, and he knows it. Much of the innovation David brings to Hearst is around cost-cutting to return cash to the mother ship: for example, putting one editor in charge of multiple titles, leveraging the scale of the organization, recycling content through multiple channels and titles, and demonstrating discipline concerning head count.

The result? Hearst titles steal back share from digital marauders, and David rides *Cosmopolitan* (a big Hearst title) off into the sunset. Right? Well, no. Hearst Magazines will likely be a shadow of the shadow it is now in ten years. However, Hearst will be fine, as it finds and retains managers who understand the business life cycle. They know how to harvest so they can plant new trees—which they will harvest well before they become mature.

On a risk-adjusted basis, you are better off bringing an entrepreneurial mind-set to a company that has already survived its birth pains (think not A–C, but D–F). That's because the infant mortality of new tech start-ups (basically, before the Series A venture round) is greater than 75 percent. Sure, your plucky start-up might find its lane and make you rich, but it probably won't. This denial is key to our economy, as some of this crazy turns to crazy successful and fuels key parts of our economy.

Long/Short Tail

In tech, many long tails are atrophying. Take digital advertising, for example. Facebook and Google accounted for 90 percent of U.S. digital advertising revenue growth in 2016. You are better off picking (if possible) one of a handful of winners (Google/FB/MSFT) or firms in their ecosystem. Disruptors that break open new markets are rare—lottery winners.

In some traditional consumer goods industries, however, the long tail is growing. Thus, it's better to work for Google than a niche search player; but conversely, it's better to work for a craft brewery

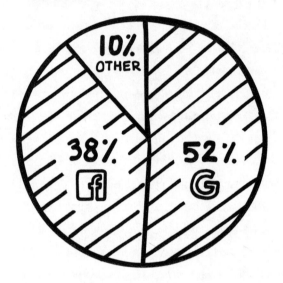

DIGITAL AD GROWTH
2016

10%
OTHER

38%

52%

Kint, Jason. "Google and Facebook Devour the Ad and Data Pie. Scraps for Everyone Else." Digital Content Next.

than Miller. The very concentration of tech space into dominant information platforms (Amazon reviews, Google, Trip Advisor) has facilitated the identification of breakout nontech products from unknown makers, and the niche-ification of traditional categories. Small players can get global reach and instant credibility without the massive ad budgets and distribution networks that their larger competitors once used to limit the market. The long tail has new life in consumer, as discretionary income wants special, not big.

We are seeing this across categories. In cosmetics, for example, brands including NYX and Anastasia Beverly Hills are challenging the traditional giants by going straight to influencers on Instagram and other social platforms, and responding to trends, registered on

Google, with a supply chain that gets products to market in a fraction of the time of traditional players. The result is that they are getting many times the brand exposure of their traditional competitors, with fractions of their ad spend. For example, with less than 1 percent of L'Oréal's Google keyword purchases, NYX has five times L'Oréal's organic visibility. In sporting goods, niche players in categories such as skis, mountain bikes, and running shoes are grabbing chunks of the high-margin enthusiast market by signing up young influencers, deft online promotion, and uber-fast product introductions.

The Myth of Balance

There are people who are successful professionally while managing a food blog, volunteering at the animal shelter, and mastering ballroom dance. Assume you are not one of those people. Balance is largely a myth when establishing your career. The slope of the trajectory for your career is (unfairly) set the first five years postgraduation. If you want the trajectory to be steep, you'll need to burn a lot of fuel. The world is not yours for the taking, but for the trying. Try hard, really hard.

I have a lot of balance now. That's a function of the lack of balance in my twenties and thirties. Other than business school, from twenty-two to thirty-four, I remember work and not much else. The world does not belong to the big, but to the fast. You want to cover more ground in less time than your peers. This is partially talent, but mostly endurance. My lack of balance as a young professional cost me my hair, my first marriage, and arguably my twenties. And it was worth it.

Are You an Entrepreneur?

I began this chapter describing some of the characteristics I see across the board in successful people in the digital age. But along our varied, digital-age career path, many people will at some point consider an entrepreneurial opportunity, whether it be starting their own business, joining an existing start-up, or launching a new business with a larger organization.

This is a good thing—new ventures are important for injecting new energy and ideas into the economy, and are a major source of wealth creation for those lucky and smart enough to be involved in a firm that defies the odds and prospers. Billionaire founders from Sam Walton to Mark Zuckerberg are familiar characters in business lore, and successes can create whole tribes of wealthy people overnight. The "Microsoft millionaire" is a cultural touchstone in the Seattle area, where one economist estimates the company created 10,000 millionaires by 2000.

Culturally, we have elevated the entrepreneur to iconic status, along with sports heroes and entertainment stars. It's a fundamental American myth, from Ayn Rand's still influential personification of entrepreneurial independence in Hank Rearden to the mythmaking that erupted upon the death of Steve Jobs. Entrepreneurs are seen as individual, self-made visionaries with vast wealth. They are perhaps the purest expression of the American hero. Superhero, even. Superman can reverse the rotation of the Earth, but Iron Man Tony Stark would be better on an earnings call and is a very human superhero—Elon Musk.

As we've discussed, it's not for most people—and the odds against you seem heavier by the year. In fact, very few people have the personality characteristics and skills that make up a successful entrepreneur. And it isn't about being "good enough" or "smart enough"—indeed, some of the characteristics of successful entrepreneurs are real detriments in other aspects of life.

So, how do you know if you're an entrepreneur?

The traits of successful entrepreneurs haven't changed much in the digital age: you need more builders than branders, and it's key to have a technologist as part of, or near, the founding team. There are three tests or questions:

1. Are you comfortable with public failure?

2. Do you like to sell?

3. Do you lack the skills to work at a big firm?

I know people who have all the skills to build great businesses. But they'll never do so, because they could never go to work only to, at the end of the month, in exchange for working eighty hours a week, write the firm a check.

Unless you have built firms and shepherded them to successful exits, or have access to seed capital (most don't, and it's always expensive), then you'll need to pay the company for the right to work your ass off until you can raise money. And most start-ups never raise the needed money. Most people can't wrap their head around the notion of working without getting paid—and 99 percent plus will never risk their own capital for the pleasure of . . . working.

Are you comfortable with public failure?

Most failures are private: you decide law school isn't for you (bombed the LSAT); you decide to spend more time with your kids (got fired), or you're working on a bunch of "projects" (can't get a job). However, there's no hiding with your own business failure. It's *you*, and if you're so awesome, it must succeed . . . right? Wrong, and when it doesn't, it feels like elementary school, where the marketplace is a sixth grader laughing at you because you've wet your pants . . . times a hundred.

Do you like to sell?

"Entrepreneur" is a synonym for "salesperson." Selling people to join your firm, selling them to stay at your firm, selling investors, and (oh yeah) selling customers. It doesn't matter if you're running the corner store or Pinterest, if you plan to start a business, you'd better be damn good at selling. Selling is calling people who don't want to hear from you, pretending to like them, getting treated poorly, and then calling them again. I likely won't start another business because my ego is getting too big (and my intestinal fortitude too weak) to sell.

I, incorrectly, believe that our collective genius at L2 should mean the product sells itself—and sometimes it does. There has to be a product that doesn't require you to get out the spoon and publicly eat shit over and over. Actually, no, there isn't.

Google has an algorithm that can answer anything and identify people who have explicitly declared an interest in buying your product, then advertise to those people at that exact moment. Yet Google still has to hire thousands of attractive people with average IQs and exceptional EQ to sell the shit out of . . . Google. Entrepreneurship is a sales job with negative commissions for the first three to five years, or until you go out of business—whichever comes first.

The good news: if you like to sell and you're good at it, you will always make more money, relative to how hard you work, than any of your colleagues, and . . . they will hate you for it.

Do you lack the skills to work at a big firm?

Being successful in a big firm is not easy and requires a unique skill set. You have to play nice with others, suffer injustices and bullshit at every turn, and be politically savvy—get noticed by key stakeholders doing good work and garner executive-level sponsorship. However, if you are good at working at a big firm, then, on a risk-adjusted basis, you are better off doing just that—and not struggling against the long odds of working at a small firm. Big firms are great platforms that can scale your skills.

If, on the other hand, you don't play well with others, have an inability to trust your fate to others, and are almost clinically obsessed with your vision for a new product or service, you may be an entrepreneur. I know I am one: past potential employers viewed me as such an asshole I had to start my own business. For me, entrepreneurship was a survival mechanism, as I didn't have the skills to be successful at the greatest platforms in history, big U.S. companies.

At small firms, the highs are very high and the lows even lower. My greatest joy and pride are my kids. Second are the firms we've started, even the failures. As with kids, there is an instinctive, genetic connection with a firm you've founded. It looks, smells, and feels like you, and you can't help but feel joy and pride when it takes its first step. And it's like your child coming home with a good report card when that company is written up as one of the fastest-growing firms in Ronkonkoma, New York.

Just as important, and unlike parenthood, most people deep down know they could never do what you do. There's an admiration for entrepreneurs, as they are the engine of job growth and uniquely American in their optimism and willingness to take risks.

That said, in our digital age, with the endless and well-publicized stories of billionaire college dropouts, we idealize entrepreneurship. Ask yourself, and some people you trust, the above questions about your personality and skills. If you answer positively on the first two, and you are not skilled at working at a big company, then step into the cage of chaos monkeys.

Chapter 11

After the Horsemen

In a democratic society the existence of large centers of private power is dangerous to the continuing vitality of a free people.

—Louis Brandeis

The Four manifest god, love, sex, and consumption and add value to billions of people's lives each day. However, these firms are not concerned with the condition of our souls, will not take care of us in our old age, nor hold our hand. They are organizations that have aggregated enormous power. Power corrupts, especially in a society infected with what the pope calls the "idolatry of money." These companies avoid taxes, invade privacy, and destroy jobs to increase profits because . . . they can. The concern is not only that firms do this, but that the Four have become so good at it.

It took Facebook less than a decade to reach 1 billion customers. Now it's a global communications utility, with a nose to becoming the world's biggest advertising company. It's a company with 17,000 employees valued at \$448 billion.[1,2] The riches flow to the lucky few. Disney, a hugely successful media company by traditional standards, commands less than half that market capitalization (\$181 billion), but employs 185,000 people.[3,4]

This uber-productivity creates growth, but not necessarily prosperity. Giants of the industrial age, including General Motors and IBM, employed hundreds of thousands of workers. The spoils were carved up more fairly than today. Investors and executives got rich, though not billionaires, and workers, many of them unionized, could buy homes and motorboats and send their kids to college.

That's the America that millions of angry voters want back. They tend to blame global trade and immigrants, but the tech economy, and its fetishization, is as much to blame. It has dumped an enormous amount of wealth into the laps of a small cohort of investors and incredibly talented workers—leaving much of the workforce behind (perhaps believing the opiate of the masses will be streaming video content and a crazy-powerful phone).

Together the horsemen employ about 418,000 employees—the population of Minneapolis.[5] If you combine the value of the Four Horsemen's public shares of stock, it comes to \$2.3 *trillion*.[6] That means our 2.0 version of Minneapolis contains nearly as much wealth as the gross domestic product of France, a developed nation of 67 million citizens.[7] This affluent city will thrive while all the rest of Minnesota scrounges for investment, opportunity, and jobs.

This reckoning is happening. It's the distortion created by the steady march of digital technology, the dominance of the Four, and a belief that the "innovator" class deserves an exponentially better life.

It's dangerous for society, and it shows no sign of slowing down. It hollows out the middle class, which leads to bankrupt towns, feeds the angry politics of those who feel cheated, and underpins the rise of demagogues. I'm not a policy expert, and I won't weigh down this book with a lot of prescriptions I'm not qualified to make. However, the distortions are visible and disturbing.

Purpose

How are we using our brain power, and to what purpose? Think back to the middle of the twentieth century. When it came to computing power, we were impoverished. Computers were big primitive tabulators, with transistors gradually replacing vacuum tubes. There was no artificial intelligence, and search took place at a snail's pace, in libraries, through something called a card catalogue.

Despite those handicaps, we tackled huge projects for humanity. First, there was the race to save the world, and split the atom. Hitler had a head start, and if the Nazis got there first, it would have been game over. In 1939, the U.S. government launched the Manhattan Project. Within six years, some 130,000 people were mobilized. That's about a third of Amazon's workforce.

Within six years, we had won the race to the bomb. You may not look at that as a worthy goal. But it was a strategic priority to win that technology race, and we mobilized to do it. We did the same thing to reach the Moon, an endeavor that, at its peak, involved 400,000 workers from the United States, Canada, and Britain.

Each of the horsemen dwarfs both the Manhattan and Apollo projects in intelligence and technological capacity. Their computing power is near limitless, and ridiculously cheap. They inherit three generations of research on statistical analysis, optimization, and artificial intelligence. Each horseman swims in data we hemorrhage 24/7, analyzed by some of the most intelligent, creative, and determined people who have ever lived.

What is the endgame for this, the greatest concentration of human and financial capital ever assembled? What is their mission? Cure cancer? Eliminate poverty? Explore the universe? No, their goal: to sell another fucking Nissan.

The heroes and innovators of yesteryear created, and still create, jobs for hundreds of thousands of people. Unilever has a $156 billion market cap spread over 171,000 middle-class households.[8,9] Intel has a $165 billion market cap and employs 107,000 people.[10,11] Compare that to Facebook, which has a $448 billion market cap and 17,000 employees.[12,13]

We have a perception of these large companies that they must be creating a lot of jobs, but in fact they have a small number of high-paying jobs, and everybody else is fighting over the scraps. America is on pace to be home to 3 million lords and 350 million serfs. Again, it's never been easier to be a billionaire, but never been harder to be a millionaire.

It may be futile, or just wrong, to fight them or blanket-label these incredible firms as "bad." I don't know. However, I am certain that understanding the Four gives insight into our digital age and a greater capacity to build economic security for you and your family. I hope this book helps with both.

Acknowledgments

I'm (really) glad this whole first book thing is done, but hoping we can keep the team together. My agent, Jim Levine, is great at what he does (I knew that). The bonus is he's become a role model: married fifty years, smart and strong. This book is as much his as mine. My editor, Niki Papadopoulos, kept the work honest and on deadline.

My partners at L2, Maureen Mullen and Katherine Dillon, have been a constant source of inspiration and camaraderie. I hope they take pride in this work, as they shaped it. L2's CEO, Ken Allard, has been hugely supportive and generous. Several outstanding professionals at L2 informed the thinking in this book:

—Danielle Bailey

—Todd Benson (board)

—Colin Gilbert

—Claude de Jocas

—Mabel McClean

Acknowledgments

The book team at L2, Elizabeth Elder, Ariel Meranus, Maria Petrova, and Kyle Scallon, took lemonade and made it much better (like lemonade and . . . vodka). NYU Stern colleagues Adam Brandenburger, Anastasia Crosswhite, Vasant Dhar, Peter Henry, Elizabeth Morrison, Rika Nazem, and Luke Williams have been supportive and tolerant.

I'd also like to thank my parents for having the courage to get on a steamship to America, and California taxpayers and the Regents of the University of California for giving an unremarkable kid remarkable opportunities.

Beata, thanks, and I love you.

Illustration Credits

Market Capitalization, as of April 25, 2017

Yahoo! Finance. https://finance.yahoo.com/

Return on Human Capital, 2016

Forbes, May, 2016. https://www.forbes.com/companies/general-motors/
Facebook, Inc. https://newsroom.fb.com/company-info/
Yahoo! Finance. https://finance.yahoo.com/

The Five Largest Companies, in 2006

Taplin, Jonathan. "Is It Time to Break Up Google?" *The New York Times*.

Where People Start Product Searches, 2016

Soper, Spencer. "More Than 50% of Shoppers Turn First to Amazon in Product Search." Bloomberg.

Percent of American Households Using Amazon Prime, 2016

"Sizeable Gender Differences in Support of Bans on Assault Weapons, Large Clips." Pew Research Center.
ACTA, "The Vote Is In—78 Percent of U.S. Households Will Display Christmas Trees This Season: No Recount Necessary Says American Christmas Tree Association." ACTA.

Illustration Credits

"2016 November General Election Turnout Rates." United States Elections Project.

Stoffel, Brian. "The Average American Household's Income: Where Do You Stand?" *The Motley Fool.*

Green, Emma. "It's Hard to Go to Church." *The Atlantic.*

"Twenty Percent of U.S. Households View Landline Telephones as an Important Communication Choice." The Rand Corporation.

Tuttle, Brad. "Amazon Has Upper-Income Americans Wrapped Around Its Finger." *Time.*

Flash Sale Sites' Industry Revenue

Lindsey, Kelsey. "Why the Flash Sale Boom May Be Over—And What's Next." RetailDIVE.

2006–2016 Stock Price Growth

Choudhury, Mawdud. "Brick & Mortar U.S. Retailer Market Value—2006 Vs Present Day." ExecTech.

Stock Price Change on 1/5/2017

Yahoo! Finance. https://finance.yahoo.com/

U.S. Market Shares, Apparel & Accessories

Peterson, Hayley. "Amazon Is About to Become the Biggest Clothing Retailer in the US." *Business Insider.*

Average Monthly Spend on Amazon, U.S. Average 2016

Shi, Audrey. "Amazon Prime Members Now Outnumber Non-Prime Customers." *Fortune.*

Percentage of Affluents Who Can Identify a "Favorite Brand"

Findings from the 10th Annual Time Inc./YouGov Survey of Affluence and Wealth, April 2015.

Industry Value in the U.S.

Farfan, Barbara. "2016 US Retail Industry Overview." The Balance.

"Value of the Entertainment and Media Market in the United States from 2011 to 2020 (in Billion U.S. Dollars)." Statista.

"Telecommunications Business Statistics Analysis, Business and Industry Statistics." Plunkett Research.

U.S. Retail Employees

"Retail Trade." DATAUSA.

The Smartphone Global Marketshare vs. Profits, 2016
Sumra, Husain. "Apple Captured 79% of Global Smartphone Profits in 2016." MacRumors.

Gap vs. Levi's: Revenue in Billions
Gap Inc., Form 10-K for the Period Ending January 31, 1998 (filed March 13, 1998), from Gap, Inc. website.

Gap Inc., Form 10-K for the Period Ending January 31, 1998 (filed March 28, 2006), from Gap, Inc. website.

"Levi Strauss & Company Corporate Profile and Case Material." Clean Clothes Campaign.

Levi Strauss & Co., Form 10-K for the Period Ending November 27, 2005 (filed February 14, 2006), p. 26, from Levi Strauss & Co. website.

Cost of College
"Do you hear that? It might be the growing sounds of pocketbooks snapping shut and the chickens coming home . . ." AEIdeas, August 2016. http://bit.ly/2nHvdfir.

Irrational Exuberance, Robert Shiller. http://amzn.to/2o98DZE.

Time Spent on Facebook, Instagram, & WhatsApp per Day, December 2016
"How Much Time Do People Spend on Social Media?" MediaKix.

Number of Timeline Posts per Day—Single vs. In a Relationship
Meyer, Robinson. "When You Fall in Love This Is What Facebook Sees." *The Atlantic.*

Individuals Moving from/to WPP to Facebook & Google
L2 Analysis of LinkedIn Data.

Global Reach vs. Engagement by Platform
L2 Analysis of Unmetric Data.

L2 Intelligence Report: Social Platforms 2017. L2, Inc.

U.S. Digital Advertising Growth, 2016 YOY
Kafka, Peter. "Google and Facebook are booming. Is the rest of the digital ad business sinking?" *Recode.*

Market Capitalization, February 2016
Yahoo! Finance. Accessed in February 2016. https://finance.yahoo.com/

Illustration Credits

YOY Performance of Top CPG Brands, 2014–2015

"A Tough Road to Growth: The 2015 Mid-Year Review: How the Top 100 CPG Brands Performed." Catalina Marketing.

Percent of Global Revenue Outside the U.S., 2016

"Facebook Users in the World." Internet World Stats.

"Facebook's Average Revenue Per User as of 4th Quarter 2016, by Region (in U.S. Dollars)." Statista.

Millward, Steven. "Asia Is Now Facebook's Biggest Region." Tech in Asia.

Thomas, Daniel. "Amazon Steps Up European Expansion Plans." *The Financial Times.*

Alibaba.com, YOY Growth, 2014–2016

Alibaba Group, FY16-Q3 for the Period Ending December 31, 2016 (filed January 24, 2017), p. 2, from Alibaba Group website.

Price:Sales Ratio, April 28, 2017

Yahoo! Finance. https://finance.yahoo.com/.

LinkedIn Revenue Sources, 2015

LinkedIn Corporate Communications Team. "LinkedIn Announces Fourth Quarter and Full Year 2015 Results." LinkedIn.

The March to a Billion Users

Desjardins, Jeff. "Timeline: The March to a Billion Users [Chart]." Visual Capitalist.

Digital Ad Growth, 2016

Kint, Jason. "Google and Facebook Devour the Ad and Data Pie. Scraps for Everyone Else." Digital Content Next.

Notes

Chapter 1: The Four

1. Zaroban, Stefany. "US e-commerce sales grow 15.6% in 2016." Digital Commerce 360. February 17, 2017. https://www.digitalcommerce360.com/2017/02/17/us-e-commerce-sales-grow-156-2016/.

2. "2017 Top 250 Global Powers of Retailing." National Retail Federation. January 16, 2017. https://nrf.com/news/2017-top-250-global-powers-of-retailing.

3. Yahoo! Finance. https://finance.yahoo.com/.

4. "The World's Billionaires." *Forbes*. March 20, 2017. https://www.forbes.com/billionaires/list/.

5. Amazon.com, Inc., FY16-Q4 for the Period Ending December 31, 2016 (filed February 2, 2017), p. 13, from Amazon.com, Inc. website. http://phx.corporate-ir.net/phoenix.zhtml?c=97664&p=irol-reportsother.

6. "Here Are the 10 Most Profitable Companies." *Forbes*. June 8, 2016. http://fortune.com/2016/06/08/fortune-500-most-profitable-companies-2016/.

7. Miglani, Jitender. "Amazon vs Walmart Revenues and Profits 1995-2014." Revenues and Profits. July 25, 2015. https://revenuesandprofits.com/amazon-vs-walmart-revenues-and-profits-1995-2014/.

8. FY16-Q4 for the Period Ending December 31, 2016.

9. "Apple Reports Fourth Quarter Results." Apple Inc. October 25, 2016. http://www.apple.com/newsroom/2016/10/apple-reports-fourth-quarter-results.html.

10. Wang, Christine. "Apple's cash hoard swells to record $246.09 billion." CNBC. January 31, 2017. http://www.cnbc.com/2017/01/31/apples-cash-hoard-swells-to-record-24609-billion.html.

11. "Denmark GDP 1960-2017." Trading Economics. 2017. http://www.tradingeconomics.com/denmark/gdp.

12. "Current World Population." Worldometers. April 25, 2017.

13. Facebook, Inc. https://newsroom.fb.com/company-info/.

14. Ng, Alfred. "Facebook, Google top out most popular apps in 2016." CNET. December 28, 2016. https://www.cnet.com/news/facebook-google-top-out-uss-most-popular-apps-in-2016/.

15. Stewart, James B. "Facebook Has 50 Minutes of Your Time Each Day. It Wants More." New York Times. May 5, 2016. https://www.nytimes.com/2016/05/06/business/facebook-bends-the-rules-of-audience-engagement-to-its-advantage.html?_r=0.

16. Lella, Adam, and Andrew Lipsman. "2016 U.S. Cross-Platform Future in Focus." comScore. March 30, 2016. https://www.comscore.com/Insights/Presentations-and-Whitepapers/2016/2016-US-Cross-Platform-Future-in-Focus.

17. Ghoshal, Abhimanyu. "How Google handles search queries it's never seen before." The Next Web. October 26, 2015. https://thenextweb.com/google/2015/10/26/how-google-handles-search-queries-its-never-seen-before/#.tnw_Ma3rOqjl.

18. "Alphabet Announces Third Quarter 2016 Results." Alphabet Inc. October 27, 2016. https://abc.xyz/investor/news/earnings/2016/Q3_alphabet_earnings/.

19. Lardinois, Frederic. "Google says there are now 2 billion active Chrome installs." TechCrunch. November 10, 2016. https://techcrunch.com/2016/11/10/google-says-there-are-now-2-billion-active-chrome-installs/.

20. Forbes. May, 2016. https://www.forbes.com/companies/general-motors/.

21. Facebook, Inc. https://newsroom.fb.com/company-info/.

22. Yahoo! Finance. https://finance.yahoo.com/.

23. Ibid.

24. "Report for Selected Countries and Subjects." International Monetary Fund. October, 2016. http://bit.ly/2eLOnMI.

25. Soper, Spencer. "More Than 50% of Shoppers Turn First to Amazon in Product Search." Bloomberg. September 27, 2016. https://www.bloomberg.com

/news/articles/2016-09-27/more-than-50-of-shoppers-turn-first-to-amazon
-in-product-search.

Chapter 2: Amazon

1. "Sizeable gender differences in support of bans on assault weapons, large clips." Pew Research Center. August 9–16, 2016. http://www.people-press .org/2016/08/26/opinions-on-gun-policy-and-the-2016-campaign/august guns_6/.

2. Ibid.

3. Gajanan, Mahita. "More Than Half of the Internet's Sales Growth Now Comes From Amazon." *Fortune.* February 1, 2017. http://fortune.com/2017/02/01/am azon-online-sales-growth-2016/.

4. Amazon. 2016 Annual Report. February 10, 2017. http://phx.corporate-ir.net /phoenix.zhtml?c=97664&p=irol-sec&control_selectgroup=Annual%20Fil ings#14806946.

5. "US Retail Sales, Q1 2016-Q4 2017 (trillions and % change vs. same quarter of prior year)." eMarketer. February 2017. http://dashboard-na1.emarketer.com /numbers/dist/index.html#/584b26021403070290f93a2d/5851918a0626310a 2c186ac2.

6. Weise, Elizabeth. "That review you wrote on Amazon? Priceless." *USA Today.* March 20, 2017. https://www.usatoday.com/story/tech/news/2017/03/20/re view-you-wrote-amazon-priceless/99332602/.

7. Kim, Eugene. "This Chart Shows How Amazon Could Become the First $1 Trillion Company." *Business Insider.* December 7, 2016. http://www.busines sinsider.com/how-amazon-could-become-the-first-1-trillion-business -2016-12.

8. *The Cambridge Encyclopedia of Hunters and Gatherers.* Edited by Richard B. Lee and Richard Daly (Cambridge University Press: 2004). "Introduction: Foreigners and Others."

9. Taylor, Steve. "Why Men Don't Like Shopping and (Most) Women Do: The Origins of Our Attitudes Toward Shopping." *Psychology Today.* February 14, 2014. https://www.psychologytoday.com/blog/out-the-darkness/201402/why -men-dont-shopping-and-most-women-do.

10. "Hunter gatherer brains make men and women see things differently." *Telegraph.* July 30, 2009. http://www.telegraph.co.uk/news/uknews/5934226 /Hunter-gatherer-brains-make-men-and-women-see-things-differently .html.

11. Van Aswegen, Anneke. "Women vs. Men—Gender Differences in Purchase Decision Making." *Guided Selling.* October 29, 2015. http://www.guided-selling .org/women-vs-men-gender-differences-in-purchase-decision-making.

12. Duenwald, Mary. "The Psychology of . . . Hoarding." *Discover.* October 1, 2004. http://discovermagazine.com/2004/oct/psychology-of-hoarding.

13. "Number of Americans with Diabetes Projected to Double or Triple by 2050." Centers for Disease Control and Prevention. October 22, 2010. https://www .cdc.gov/media/pressrel/2010/r101022.html.

14. "Paul Pressler Discusses the Impact of Terrorist Attacks on Theme Park Industry." CNN.com/Transcripts. October 6, 2001. http://transcripts.cnn.com /TRANSCRIPTS/0110/06/smn.26.html.

15. "Euro rich list: The 48 richest people in Europe." *New European.* February 26, 2017. http://www.theneweuropean.co.uk/culture/euro-rich-list-the-48-richest -people-in-europe-1-4906517.

16. "LVMH: Luxury's Global Talent Academy." *The Business of Fashion.* April 25, 2017. https://www.businessoffashion.com/community/companies/lvmh.

17. Fernando, Jason. "Home Depot Vs. Lowes: The Home Improvement Battle." Investopedia. July 7, 2015.

18. Bleakly, Fred R. "The 10 Super Stocks of 1982." *New York Times.* January 2, 1983. http://www.nytimes.com/1983/01/02/business/the-10-super-stocks-of -1982.html?pagewanted=all.

19. Friedman, Josh. "Decade's Hottest Stocks Reflect Hunger for Anything Tech." *Los Angeles Times.* December 28, 1999. http://articles.latimes.com/1999/dec /28/business/fi-48388.

20. Recht, Milton. "Changes in the Top Ten US Retailers from 1990 to 2012: Six of the Top Ten Have Been Replaced." *Misunderstood Finance.* October 21, 2013. http://misunderstoodfinance.blogspot.com.co/2013/10/changes-in-top -ten-us-retailers-from.html.

21. Farfan, Barbara. "Largest US Retail Companies on 2016 World's Biggest Retail Chains List." The Balance. February 13, 2017. https://www.thebalance.com /largest-us-retailers-4045123.

22. Kim, Eugene. "Amazon Sinks on Revenue Miss." *Business Insider.* February 2, 2017. http://www.businessinsider.com/amazon-earnings-q4-2016-2017-2.

23. Miglani, Jitender. "Amazon vs Walmart Revenues and Profits 1995-2014." July 25, 2015. Revenues and Profits. http://revenuesandprofits.com/amazon-vs -walmart-revenues-and-profits-1995-2014/.

24. Baird, Nikki. "Are Retailers Over-Promoting for Holiday 2016?" *Forbes.* December 16, 2016. https://www.forbes.com/sites/nikkibaird/2016/12/16/are-retailers-over-promoting-for-holiday-2016/#53bb6fbb3b8e.

25. Leibowitz, Josh. "How Did We Get Here? A Short History of Retail." LinkedIn. June 7, 2013. https://www.linkedin.com/pulse/20130607115409-12921524-how-did-we-get-here-a-short-history-of-retail.

26. Skorupa, Joe. "10 Oldest U.S. Retailers." *RIS.* August 19, 2008. https://risnews.com/10-oldest-us-retailers.

27. Feinberg, Richard A., and Jennifer Meoli. "A Brief History of the Mall." *Advances in Consumer Research* 18 (1991): 426–27. Acessed April 4, 2017. http://www.acrwebsite.org/volumes/7196/volumes/v18/NA-18.

28. Ho, Ky Trang. "How to Profit from the Death of Malls in America." *Forbes.* December 4, 2016. https://www.forbes.com/sites/trangho/2016/12/04/how-to-profit-from-the-death-of-malls-in-america/#7732f3cc61cf.

29. "A Timeline of the Internet and E-Retailing: Milestones of Influence and Concurrent Events." Kelley School of Business: Center for Education and Research in Retailing. https://kelley.iu.edu/CERR/timeline/print/page14868.html.

30. Nazaryan, Alexander. "How Jeff Bezos Is Hurtling Toward World Domination." *Newsweek.* July 12, 2016. http://www.newsweek.com/2016/07/22/jeff-bezos-amazon-ceo-world-domination-479508.html.

31. "Start Selling Online—Fast." Amazon.com, Inc. https://services.amazon.com/selling/benefits.htm.

32. "US Retail Sales, Q1 2016-Q4 2017." eMarketer. January 2017. http://totalaccess.emarketer.com/Chart.aspx?R=204545&dsNav=Ntk:basic%7cdepartment+of+commerce%7c1%7c,Ro:-1,N:1326,Nr:NOT(Type%3aComparative+Estimate)&kwredirect=n.

33. Del Rey, Jason. "Amazon has at least 66 million Prime members but subscriber growth may be slowing." *Recode.* February 3, 2017. https://www.recode.net/2017/2/3/14496740/amazon-prime-membership-numbers-66-million-growth-slowing.

34. Gajanan, Mahita. "More Than Half of the Internet's Sales Growth Now Comes From Amazon." *Fortune.* February 1, 2017. http://fortune.com/2017/02/01/amazon-online-sales-growth-2016/.

35. Cassar, Ken. "Two extra shopping days make 2016 the biggest holiday yet." *Slice Intelligence.* January 5, 2017. https://intelligence.slice.com/two-extra-shopping-days-make-2016-biggest-holiday-yet/.

36. Cone, Allen. "Amazon ranked most reputable company in U.S. in Harris Poll." *UPI.* February 20, 2017. http://www.upi.com/Top_News/US/2017/02/20/Ama zon-ranked-most-reputable-company-in-US-in-Harris-Poll/6791487617347/.

37. "Amazon's Robot Workforce Has Increased by 50 Percent." CEB Inc. December 29, 2016. https://www.cebglobal.com/talentdaily/amazons-robot-workforce -has-increased-by-50-percent/.

38. Takala, Rudy. "Top 2 U.S. Jobs by Number Employed: Salespersons and Cash-iers." CNS News. March 25, 2015. http://www.cnsnews.com/news/article /rudy-takala/top-2-us-jobs-number-employed-salespersons-and-cashiers.

39. "Teach Trends." National Center for Education Statistics. https://nces.ed.gov /fastfacts/display.asp?id=28.

40. Full transcript: Internet Archive founder Brewster Kahle on Recode Decode. *Recode.* March 8, 2017. https://www.recode.net/2017/3/8/14843408/transcript -internet-archive-founder-brewster-kahle-wayback-machine-recode-decode.

41. Amazon Dash is a button you place anywhere in your home that connects to the Amazon app through Wi-Fi for one-click ordering. https://www.amazon .com/Dash-Buttons/b?ie=UTF8&node=10667898011.

42. http://www.businessinsider.com/amazon-prime-wardrobe-2017-6.

43. Daly, Patricia A. "Agricultural employment: Has the decline ended?" Bureau of Labor Statistics. November 1981. https://stats.bls.gov/opub/mlr/1981/11 /art2full.pdf.

44. Hansell, Saul. "Listen Up! It's Time for a Profit; A Front-Row Seat as Amazon Gets Serious." *New York Times.* May 20, 2001. http://www.nytimes.com /2001/05/20/business/listen-up-it-s-time-for-a-profit-a-front-row-seat-as-ama zon-gets-serious.html.

45. Yahoo! Finance. https://finance.yahoo.com/.

46. Damodaran, Aswath. "Enterprise Value Multiples by Sector (US)." NYU Stern. January 2017. http://pages.stern.nyu.edu/~adamodar/New_Home_Page/data file/vebitda.html.

47. Nelson, Brian. "Amazon Is Simply an Amazing Company." Seeking Alpha. December 6, 2016. https://seekingalpha.com/article/4028547-amazon-simply -amazing-company.

48. "Wal-Mart Stores' (WMT) CEO Doug McMillon on Q1 2016 Results— Earnings Call Transcript." Seeking Alpha. May 19, 2015. https://seekingalpha .com/article/3195726-wal-mart-stores-wmt-ceo-doug-mcmillon-on-q1-2016 -results-earnings-call-transcript?part=single.

49. Rego, Matt. "Why Walmart's Stock Price Keeps Falling (WMT)." Seeking Alpha. November 11, 2015. http://www.investopedia.com/articles/markets/111115 /why-walmarts-stock-price-keeps-falling.asp.

50. Rosoff, Matt. "Jeff Bezos: There are 2 types of decisions to make, and don't confuse them." *Business Insider.* April 5, 2016. http://www.businessinsider.com /jeff-bezos-on-type-1-and-type-2-decisions-2016-4.

51. Amazon.com. 2016 Letter to Shareholders. Accessed April 25, 2017. http:// phx.corporate-ir.net/phoenix.zhtml?c=97664&p=irol-reportsannual.

52. Bishop, Todd. "The cost of convenience: Amazon's shipping losses top $7B for first time." GeekWire. February 9, 2017. http://www.geekwire.com/2017 /true-cost-convenience-amazons-annual-shipping-losses-top-7b-first -time/.

53. Letter to Shareholders.

54. Stanger, Melissa, Emmie Martin, and Tanza Loudenback. "The 50 richest people on earth." *Business Insider.* January 26, 2016. http://www.businessinsider .com/50-richest-people-on-earth-2016-1.

55. "The Global Unicorn Club." *CB Insights.* https://www.cbinsights.com/research -unicorn-companies.

56. Amazon.com. FY16-Q4 for the Period Ending December 31, 2016 (filed February 2, 2017), p. 13, from Amazon.com, Inc. website. http://phx.corporate-ir .net/phoenix.zhtml?c=97664&p=irol-reportsother.

57. Goodkind, Nicole. "Amazon Beats Apple as Most Trusted Company in U.S.: Harris Poll." Yahoo! Finance. February 12, 2013. http://finance.yahoo.com /blogs/daily-ticker/amazon-beats-apple-most-trusted-company-u-harris -133107001.html.

58. Adams, Susan. "America's Most Reputable Companies, 2015." *Forbes.* May 13, 2015. https://www.forbes.com/sites/susanadams/2015/05/13/americas-most- reputable-companies-2015/#4b231fd21bb6.

59. Dignan, Larry. "Amazon posts its first net profit." CNET. February 22, 2002. https://www.cnet.com/news/amazon-posts-its-first-net-profit/.

60. Amazon.com. 2015 Q1-Q3 Quarterly Reports. Accessed April 7, 2017. http:// phx.corporate-ir.net/phoenix.zhtml?c=97664&p=irol-sec&control_select group=Quarterly%20Filings#10368189.

61. King, Hope. "Amazon's $160 billion business you've never heard of." CNN Tech. November 4, 2015. http://money.cnn.com/2015/11/04/technology/amazon-aws -160-billion-dollars/.

62. http://www.marketwatch.com/investing/stock/twtr/financials.

63. L2 Inc. "Scott Galloway: This Is the Top of the Market." L2 Inc. February 16, 2017. https://www.youtube.com/watch?v=uIXJNt-7aY4&t=1m8s.

64. https://www.nytimes.com/2017/06/16/business/dealbook/amazon-whole -foods.html?_r=0.

65. Rao, Leena. "Amazon Prime Now Has 80 Million Members." *Fortune*. April 25, 2017. http://fortune.com/2017/04/25/amazon-prime-growing-fast/.

66. Griffin, Justin. "Have a look inside the 1-million-square-foot Amazon fulfill-ment center in Ruskin." *Tampa Bay Times*. March 30, 2016. http://www.tam pabay.com/news/business/retail/have-a-look-inside-the-1-million-square -foot-amazon-fulfillment-center-in/2271254.

67. Tarantola, Andrew. "Amazon is getting into the oceanic freight shipping game." *Engadget*. January 14, 2016. https://www.engadget.com/2016/01/14/amazon -is-getting-into-the-oceanic-freight-shipping-game/.

68. Ibid.

69. Yahoo! Finance. https://finance.yahoo.com/.

70. Kapner, Suzanne. "Upscale Shopping Centers Nudge Out Down-Market Malls." *Wall Street Journal*. April 20, 2016. https://www.wsj.com/articles/upscale -shopping-centers-nudge-out-down-market-malls-1461193411?ru=yahoo? mod=yahoo_itp.

71. https://www.nytimes.com/2017/06/16/business/dealbook/amazon-whole-foods .html?_r=0.

72. https://www.nytimes.com/2017/06/16/business/dealbook/amazon-whole -foods.html?_r=0.

73. https://www.recode.net/2017/3/8/14850324/amazon-books-store-bellevue -mall-expansion.

74. Addady, Michal. "Here's How Many Pop-Up Stores Amazon Plans to Open." *Fortune*. September 9, 2016. http://fortune.com/2016/09/09/amazon-pop-up -stores/.

75. Carrig, David. "Sears, J.C. Penney, Kmart, Macy's: These retailers are closing stores in 2017." *USA Today*. May 9, 2017. https://www.usatoday.com/story /money/2017/03/22/retailers-closing-stores-sears-kmart-jcpenney-macys -mcsports-gandermountian/99492180/.

76. http://clark.com/shopping-retail/confirmed-jcpenney-stores-closing/.

77. WhatIs.com. "Bom File Format." http://whatis.techtarget.com/fileformat /BOM-Bill-of-materials-file.

78. Coster, Helen. "Diapers.com Rocks Online Retailing." *Forbes.* April 8, 2010. https://www.forbes.com/forbes/2010/0426/entrepreneurs-baby-diapers-e-commerce-retail-mother-lode.html.

79. Wauters, Robin. "Confirmed: Amazon Spends $545 Million on Diapers.com Parent Quidsi." *TechCrunch.* November 8, 2010. https://techcrunch.com/2010/11/08/confirmed-amazon-spends-545-million-on-diapers-com-parent-quidsi/.

80. L2 Inc. "Jet.com: The $3B Hair Plugs." L2 Inc. August 9, 2016. https://www.youtube.com/watch?v=6rPEhFTFE9c.

81. Jhonsa, Eric. "Jeff Bezos' Letter Shines a Light on How Amazon Sees Itself." Seeking Alpha. April 6, 2016. https://seekingalpha.com/article/3963671-jeff-bezos-letter-shines-light-amazon-sees#alt2.

82. Boucher, Sally. "Survey of Affluence and Wealth." *WealthEngine.* May 2, 2014. https://www.wealthengine.com/resources/blogs/one-one-blog/survey-affluence-and-wealth.

83. Shi, Audrey. "Amazon Prime Members Now Outnumber Non-Prime Customers." *Fortune.* July 11, 2016. http://fortune.com/2016/07/11/amazon-prime-customers/.

84. L2 Inc. "Scott Galloway: Innovation Is a Snap." L2 Inc. October 13, 2016. https://www.youtube.com/watch?v=PhB8n-ExMck.

85. Tuttle, Brad. "Amazon Has Upper-Income Americans Wrapped Around Its Finger." *Time.* April 14, 2016. http://time.com/money/4294131/amazon-prime-rich-american-members/.

86. Holum, Travis. "Amazon's Fulfillment Costs Are Taking More of the Pie." *The Motley Fool.* December 22, 2016. https://www.fool.com/investing/2016/12/22/amazons-fulfillment-costs-are-taking-more-of-the-p.aspx.

87. L2 Inc. "Scott Galloway: Amazon Flexes." *L2 Inc.* March 3, 2016. https://www.youtube.com/watch?v=Nm7gIEKYWnc.

88. L2 Inc. "Amazon IQ: Personal Care," February 2017.

89. Kantor, Jodi, and David Streitfeld. "Inside Amazon: Wrestling Big Ideas in a Bruising Workplace." *New York Times.* August 15, 2015. https://www.nytimes.com/2015/08/16/technology/inside-amazon-wrestling-big-ideas-in-a-bruising-workplace.html?_r=1.

90. Rao, Leena. "Amazon Acquires Robot-Coordinated Order Fulfillment Company Kiva Systems for $775 Million in Cash." *TechCrunch.* March 19, 2012. https://techcrunch.com/2012/03/19/amazon-acquires-online-fulfillment-company-kiva-systems-for-775-million-in-cash/.

91. Kim, Eugene. "Amazon sinks on revenue miss." *Business Insider.* February 2, 2017. http://www.businessinsider.com/amazon-earnings-q4-2016-2017-2.

92. "Scott Galloway: Amazon Flexes."

93. Yahoo! Finance. https://finance.yahoo.com/.

94. Centre for Retail Research. "The Retail Forecast for 2017-18." Centre for Retail Research. January 24, 2017. http://www.retailresearch.org/retailforecast.php.

95. "2016 Europe 500 Report." *Digital Commerce 360.* https://www.digitalcommerce360.com/product/europe-500/#!/.

96. http://www.cnbc.com/2016/05/17/amazon-planning-second-grocery-store-report.html.

97. Amazon.com Inc. 2016 Letter to Shareholders. Accessed April 25, 2017. http://phx.corporate-ir.net/phoenix.zhtml?c=97664&p=irol-reportsannual.

98. Farfan, Barbara. "2016 US Retail Industry Overview." The Balance. August 13, 2016. https://www.thebalance.com/us-retail-industry-overview-2892699.

99. "Value of the entertainment and media market in the United States from 2011 to 2020 (in billion U.S. dollars)." Statista. https://www.statista.com/statistics/237769/value-of-the-us-entertainment-and-media-market/.

100. "Telecommunications Business Statistics Analysis, Business and Industry Statistics." Plunkett Research. https://www.plunkettresearch.com/statistics/telecommunications-market-research/.

101. https://www.nytimes.com/2017/06/16/business/dealbook/amazon-whole-foods.html?_r=0.

102. "IBISWorld Industry Report 44511: Supermarkets & Grocery Stores in the US." IBISWorld. 2017. https://www.ibisworld.com/industry-trends/market-research-reports/retail-trade/food-beverage-stores/supermarkets-grocery-stores.html.

103. Rao, Leena. "Amazon Go Debuts as a New Grocery Store Without Checkout Lines." *Fortune.* December 5, 2016. http://fortune.com/2016/12/05/amazon-go-store/.

104. https://www.nytimes.com/2017/06/16/business/dealbook/amazon-whole-foods.html?_r=0.

105. https://techcrunch.com/2017/06/17/in-wake-of-amazonwhole-foods-deal-instacart-has-a-challenging-opportunity/.

106. https://www.nytimes.com/2017/06/16/business/walmart-bonobos-merger.html?_r=0.

107. https://www.nytimes.com/2017/06/16/business/dealbook/amazon-whole-foods.html?_r=0.

108. Soper, Spencer. "More Than 50% of Shoppers Turn First to Amazon in Product Search." Bloomberg. September 27, 2016. https://www.bloomberg.com/news /articles/2016-09-27/more-than-50-of-shoppers-turn-first-to-amazon-in -product-search.

Chapter 3: Apple

1. Schmidt, Michael S., and Richard Pérez-Peña. "F.B.I. Treating San Bernardino Attack as Terrorism Case." *New York Times.* December 4, 2015. https://www .nytimes.com/2015/12/05/us/tashfeen-malik-islamic-state.html.

2. Perez, Evan, and Tim Hume. "Apple opposes judge's order to hack San Bernardino shooter's iPhone." CNN. http://www.cnn.com/2016/02/16/us/san-bernardino -shooter-phone-apple/.

3. "Views of Government's Handling of Terrorism Fall to Post-9/11 Low." Pew Research Center. December 15, 2015. http://www.people-press.org/2015/12 /15/views-of-governments-handling-of-terrorism-fall-to-post-911-low/#views -of-how-the-government-is-handling-the-terrorist-threat.

4. "Millennials: A Portrait of Generation Next." Pew Research Center. February, 2010. http://www.pewsocialtrends.org/files/2010/10/millennials-confident -connected-open-to-change.pdf.

5. "Apple: FBI seeks 'dangerous power' in fight over iPhone." *The Associated Press.* February 26, 2016. http://www.cbsnews.com/news/apple-fbi-seeks-dangerous -power-in-fight-over-iphone/.

6. Cook, Tim. "A Message to Our Customers." Apple Inc. February 16, 2016. https://www.apple.com/customer-letter/.

7. "Government's Ex Parte Application for Order Compelling Apple, Inc. to Assist Agents in Search; Memorandum of Points and Authorities; Declaration of Christopher Pluhar." United States District Court for the Central District of California. February 16, 2016. https://www.wired.com/wp-content /uploads/2016/02/SB-shooter-MOTION-seeking-asst-iPhone1.pdf.

8. Tobak, Steve. "How Jobs dodged the stock option backdating bullet." CNET. August 23, 2008. https://www.cnet.com/news/how-jobs-dodged-the-stock -option-backdating-bullet/.

9. Apple Inc., Form 10-K for the Period Ending September 26, 2015 (filed November 10, 2015), p. 24, from Apple, Inc. website. http://investor.apple.com/finan cials.cfm.

10. Gardner, Matthew, Robert S. McIntyre, and Richard Phillips. "The 35 Percent Corporate Tax Myth." Institute on Taxation and Economic Policy. March 9,

2017. http://itep.org/itep_reports/2017/03/the-35-percent-corporate-tax-myth.php#.WP5ViVPyvVp.

11. Sumra, Husain. "Apple Captured 79% of Global Smartphone Profits in 2016." *MacRumors*. March 7, 2017. https://www.macrumors.com/2017/03/07/apple-global-smartphone-profit-2016-79/.

12. "The World's Billionaires." *Forbes*. March 20, 2017. https://www.forbes.com/billionaires/list/.

13. Yarow, Jay. "How Apple Really Lost Its Lead in the '80s." *Business Insider*. December 9, 2012. http://www.businessinsider.com/how-apple-really-lost-its-lead-in-the-80s-2012-12.

14. Bunnell, David. "The Macintosh Speaks for Itself (Literally) . . ." *Cult of Mac*. May 1, 2010. http://www.cultofmac.com/40440/the-macintosh-speaks-for-itself-literally/.

15. "History of desktop publishing and digital design." Design Talkboard. http://www.designtalkboard.com/design-articles/desktoppublishing.php.

16. Burnham, David. "The Computer, the Consumer and Privacy." *New York Times*. March 4, 1984. http://www.nytimes.com/1984/03/04/weekinreview/the-computer-the-consumer-and-privacy.html.

17. Ricker, Thomas. "Apple drops 'Computer' from name." *Engadget*. January 1, 2007. https://www.engadget.com/2007/01/09/apple-drops-computer-from-name/.

18. Edwards, Jim. "Apple's iPhone 6 Faces a Big Pricing Problem Around the World." *Business Insider*. July 28, 2014. http://www.businessinsider.com/android-and-iphone-market-share-and-the-iphone-6-2014-7.

19. Price, Rob. "Apple is taking 92% of profits in the entire smartphone industry." *Business Insider*. July 13, 2015. http://www.businessinsider.com/apple-92-percent-profits-entire-smartphone-industry-q1-samsung-2015-7.

20. "Louis Vuitton Biography." *Biography*. http://www.biography.com/people/louis-vuitton-17112264.

21. Apple Newsroom. "'Designed by Apple in Calfornia' chronicles 20 years of Apple design." https://www.apple.com/newsroom/2016/11/designed-by-apple-in-california-chronicles-20-years-of-apple-design/.

22. Ibid.

23. Norman, Don. *Emotional Design: Why We Love (or Hate) Everyday Things* (New York: Basic Books, 2005).

24. Turner, Daniel. "The Secret of Apple Design." *MIT Technology Review*, May 1, 2007. https://www.technologyreview.com/s/407782/the-secret-of-apple-design/.

25. Munk, Nina. "Gap Gets It: Mickey Drexler Is Turning His Apparel Chain into a Global Brand. He wants buying a Gap T-shirt to be like buying a quart of milk. But is this business a slave to fashion?" *Fortune.* August 3, 1998. http://archive.for tune.com/magazines/fortune/fortune_archive/1998/08/03/246286/index.htm.

26. Gap Inc., Form 10-K for the Period Ending January 31, 1998 (filed March 3, 1998), from Gap, Inc. website. http://investors.gapinc.com/phoenix.zhtml?c= 111302&p=IROL-secToc&TOC=aHR0cDovL2FwaS50ZW5rd2l6YXJkLmN vbS9vdXRsaW5lLnhtbD9yZXBvPXRlbmsmaXBhZ2U9Njk0NjY1JnN1YnN pZD01Nw%3d%3d&ListAll=1.

27. Gap Inc., Form 10-K for the Period Ending January 31, 1998 (filed March 28, 2006), from Gap, Inc. website. http://investors.gapinc.com/phoenix.zhtml?c= 111302&p=IROL-secToc&TOC=aHR0cDovL2FwaS50ZW5rd2l6YXJkLmN vbS9vdXRsaW5lLnhtbD9yZXBvPXRlbmsmaXBhZ2U9NDA1NjM2O SZzdWJzaWQ9NTc%3d&ListAll=1.

28. "Levi Strauss & Company Corporate Profile and Case Material." Clean Clothes Campaign. May 1, 1998. https://archive.cleanclothes.org/news/4-companies /946-case-file-levi-strauss-a-co.html.

29. Levi Strauss & Co., Form 10-K for the Period Ending November 27, 2005 (filed February 14, 2006), p. 26, from Levi Strauss & Co. website. http://levistrauss .com/investors/sec-filings/.

30. Warkov, Rita. "Steve Jobs and Mickey Drexler: A Tale of Two Retailers." CNBC. May 22, 2012. http://www.cnbc.com/id/47520270.

31. Edwards, Cliff. "Commentary: Sorry, Steve: Here's Why Apple Stores Won't Work." Bloomberg. May 21, 2001. https://www.bloomberg.com/news/articles /2001-05-20/commentary-sorry-steve-heres-why-apple-stores-wont-work.

32. Valdez, Ed. "Why (Small) Size Matters in Retail: What Big-Box Retailers Can Learn From Small-Box Store Leaders." Seeking Alpha. April 11, 2017. https:// seekingalpha.com/article/4061817-small-size-matters-retail.

33. Farfan, Barbara. "Apple Computer Retail Stores Global Locations." The Balance. October 12, 2016. https://www.thebalance.com/apple-retail-stores-global -locations-2892925.

34. Niles, Robert. "Magic Kingdom tops 20 million in 2015 theme park atten- dance report." *Theme Park Insider.* May 25, 2016. http://www.themeparkin sider.com/flume/201605/5084.

35. Apple Inc. https://www.apple.com/shop/buy-iphone/iphone-7/4.7-inch-display -128gb-gold?afid=p238|sHVGkp8Oe-dc_mtid_1870765e38482_pcrid _138112045124_&cid=aos-us-kwgo-pla-iphone—slid—product-MN8N2LL/A.

36. http://www.techradar.com/news/phone-and-communications/mobile-phones/best-cheap-smartphones-payg-mobiles-compared-1314718.

37. Dolcourt, Jessica. "BlackBerry KeyOne keyboard phone kicks off a new Black-Berry era (hands-on)." CNET. February 25, 2017. https://www.cnet.com/prod locations ucts/blackberry-keyone/preview/.

38. Nike, Inc., Form 10-K for the Period Ending May 31, 2016 (filed July 21, 2016), p. 72, from Nike, Inc. website. http://s1.q4cdn.com/806093406/files/doc_finan cials/2016/ar/docs/nike-2016-form-10K.pdf.

39. Apple Inc., Form 10-K for the Period Ending September 24, 2016 (filed October 26, 2016), p. 43, from Apple, Inc. website. http://files.shareholder.com/down loads/AAPL/4635343320x0x913905/66363059-7FB6-4710-B4A5 -7ABFA14CF5E6/10-K_2016_9.24.2016_-_as_filed.pdf. [

40. Damodaran, Aswath. "Aging in Dog Years? The Short, Glorious Life of a Successful Tech Company!" *Musings on Markets*. December 9, 2015. http://aswath damodaran.blogspot.com/2015/12/aging-in-dog-years-short-glorious-life.html.

41. Smuts, G. L. *Lion* (Johannesburg: Macmillan South Africa: 1982), 231.

42. Dunn, Jeff. "Here's how Apple's retail business spreads across the world." *Business Insider*. February 7, 2017. http://www.businessinsider.com/apple-stores -how-many-around-world-chart-2017-2.

43. Kaplan, David. "For Retail, 'Bricks' Still Overwhelm 'Clicks' As More Than 90 Percent of Sales Happened in Stores." GeoMarketing. December 22, 2015. http://www.geomarketing.com/for-retail-bricks-still-overwhelm-clicks-as -more-than-90-percent-of-sales-happened-in-stores.

44. Fleming, Sam, and Shawn Donnan. "America's Middle-class Meltdown: Core shrinks to half of US homes." *Financial Times*. December 9, 2015. https://www. ft.com/content/98ce14ee-99a6-11e5-95c7-d47aa298f769#axzz43kCxoYVk.

45. Gates, Dominic. "Amazon lines up fleet of Boeing jets to build its own air-cargo network." *Seattle Times*. March 9, 2016. http://www.seattletimes.com /business/boeing-aerospace/amazon-to-lease-20-boeing-767s-for-its-own-air -cargo-network/.

46. Rao, Leena. "Amazon to Roll Out a Fleet of Branded Trailer Trucks." *Fortune*. December 4, 2015. http://fortune.com/2015/12/04/amazon-trucks/.

47. Stibbe, Matthew. "Google's Next Cloud Product: Google Blimps to Bring Wireless Internet to Africa." *Fortune*. June 5, 2013. https://www.forbes.com /sites/matthewstibbe/2013/06/05/googles-next-cloud-product-google-blimps -to-bring-wireless-internet-to-africa/#4439e478449b.

48. Weise, Elizabeth. "Microsoft, Facebook to lay massive undersea cable." *USA Today.* May 26, 2016. https://www.usatoday.com/story/experience/2016/05/26/microsoft-facebook-undersea-cable-google-marea-amazon/84984882/.

49. "The Nokia effect." *Economist.* August 25, 2012. http://www.economist.com/node/21560867.

50. Downie, Ryan. "Behind Nokia's 70% Drop in 10 Years (NOK)." Investopedia. September 8, 2016. http://www.investopedia.com/articles/credit-loans-mortgages/090816/behind-nokias-70-drop-10-years-nok.asp.

Chapter 4: Facebook

1. "Population of China (2017)." Population of the World. http://www.livepopulation.com/country/china.html.

2. "World's Catholic Population Grows to 1.3 Billion." *Believers Portal.* April 8, 2017. http://www.believersportal.com/worlds-catholic-population-grows-1-3-billion/.

3. Frías, Carlos. "40 fun facts for Disney World's 40th anniversary." *Statesman.* December 17, 2011. http://www.statesman.com/travel/fun-facts-for-disney-world-40th-anniversary/7ckezhCnZnB6pyiT5olyEOF/.

4. Facebook, Inc. https://newsroom.fb.com/company-info/.

5. McGowan, Tom. "Google: Getting in the face of football's 3.5 billion fans." CNN. February 27, 2015. http://edition.cnn.com/2015/02/27/football/roma-juventus-google-football/.

6. "How Much Time Do People Spend on Social Media?" Mediakix. December 15, 2016. http://mediakix.com/2016/12/how-much-time-is-spent-on-social-media-lifetime/#gs.GM2awic.

7. Stewart, James B. "Facebook Has 50 Minutes of Your Time Each Day. It Wants More." *New York Times.* May 5, 2016. https://www.nytimes.com/2016/05/06/business/facebook-bends-the-rules-of-audience-engagement-to-its-advantage.html.

8. Pallotta, Frank. "More than 111 million people watched Super Bowl LI." CNN. February 7, 2017. http://money.cnn.com/2017/02/06/media/super-bowl-ratings-patriots-falcons/.

9. Facebook, Inc. https://newsroom.fb.com/company-info/.

10. Shenk, Joshua Wolf. "What Makes Us Happy?" *Atlantic.* June 2009. https://www.theatlantic.com/magazine/archive/2009/06/what-makes-us-happy/307439/.

11. Swanson, Ana. "The science of cute: Why photos of baby animals make us happy." *Daily Herald*. September 4, 2016. http://www.dailyherald.com/arti cle/20160904/entlife/160909974/.

12. "World Crime Trends and Emerging Issues and Responses in the Field of Crime Prevention and Social Justice." UN Economic and Social Council. February 12, 2014; and UNODC, Global Study on Homicide 2013: Trends, Contexts, Data (Vienna: UNODC https://www.unodc.org/documents/data-and -analysis/statistics/crime/ECN.1520145_EN.pdf. 2013). https://www.unodc .org/unodc/en/data-and-analysis/statistics/reports-on-world-crime-trends .html.

13. Meyer, Robinson. "When You Fall in Love, This Is What Facebook Sees." *Atlantic*. February 15, 2014. http://www.theatlantic.com/technology/archive /2014/02/when-you-fall-in-love-this-is-what-facebook-sees/283865/.

14. "Number of daily active Facebook users worldwide as of 1st quarter 2017 (in millions)." Statista. https://www.statista.com/statistics/346167/facebook-global -dau/.

15. Jones, Brandon. "What Information Does Facebook Collect About Its Users?" *PSafe Blog*. November 29, 2016. http://www.psafe.com/en/blog/information -facebook-collect-users/.

16. Murphy, Mike. "Here's how to stop Facebook from listening to you on your phone." *Quartz*. June 2, 2016. https://qz.com/697923/heres-how-to-stop-face book-from-listening-to-you-on-your-phone.

17. Krantz, Matt. "13 big companies keep growing like crazy." *USA Today*. March 10, 2016. https://www.usatoday.com/story/money/markets/2016/03/10/13-big -companies-keep-growing-like-crazy/81544188/.

18. Grassegger, Hannes, and Mikael Krogerus. "The Data That Turned the World Upside Down." *Motherboard*. January 28, 2017. https://motherboard.vice.com /en_us/article/how-our-likes-helped-trump-win.

19. Cadwalladr, Carole. "Robert Mercer: The big data billionaire waging war on mainstream media." *Guardian*. February 26, 2017. https://www.theguardian .com/politics/2017/feb/26/robert-mercer-breitbart-war-on-media-steve-ban non-donald-trump-nigel-farage.

20. "As many as 48 million Twitter accounts aren't people, says study." CNBC. April 12, 2017. http://www.cnbcafrica.com/news/technology/2017/04/10/many -48-million-twitter-accounts-arent-people-says-study/.

21. L2 Analysis of LinkedIn Data.

22. Novet, Jordan. "Snapchat by the numbers: 161 million daily users in Q4 2016, users visit 18 times a day." *VentureBeat.* February 2, 2017. https://venturebeat .com/2017/02/02/snapchat-by-the-numbers-161-million-daily-users-in-q4 -2016-users-visit-18-times-a-day/.

23. Balakrishnan, Anita. "Snap closes up 44% after rollicking IPO." CNBC. March 2, 2017. http://www.cnbc.com/2017/03/02/snapchat-snap-open-trading-price -stock-ipo-first-day.html.

24. Pant, Ritu. "Visual Marketing: A Picture's Worth 60,000 Words." *Business 2 Community.* January 16, 2015. http://www.business2community.com/digital -marketing/visual-marketing-pictures-worth-60000-words-01126256#uaLl H2bk76Uj1zYA.99.

25. Khomami, Nadia, and Jamiles Lartey. "United Airlines CEO calls dragged passenger 'disruptive and belligerent.'" *Guardian.* April 11, 2017. https://www. theguardian.com/world/2017/apr/11/united-airlines-boss-oliver-munoz-says -passenger-belligerent.

26. Castillo, Michelle. "Netflix plans to spend $6 billion on new shows, blowing away all but one of its rivals." CNBC. October 17, 2016. http://www.cnbc.com /2016/10/17/netflixs-6-billion-content-budget-in-2017-makes-it-one-of-the -top-spenders.html.

27. Kafka, Peter. "Google and Facebook are booming. Is the rest of the digital ad business sinking?" *Recode.* November 2, 2016. https://www.recode.net/2016 /11/2/13497376/google-facebook-advertising-shrinking-iab-dcn.

28. Ungerleider, Neal. "Facebook Acquires Oculus VR for $2 Billion." *Fast Company.* March 25, 2014. https://www.fastcompany.com/3028244/tech-forecast /facebook-acquires-oculus-vr-for-2-billion.

29. "News companies and Facebook: Friends with benefits?" *Economist.* May 16, 2015. http://www.economist.com/news/business/21651264-facebook-and -several-news-firms-have-entered-uneasy-partnership-friends-benefits.

30. Smith, Gerry. "Facebook, Snapchat Deals Produce Meager Results for News Outlets." Bloomberg. January 24, 2017. https://www.bloomberg.com/news /articles/2017-01-24/facebook-snapchat-deals-produce-meager-results-for -news-outlets.

31. Constine, Josh. "How Facebook News Feed Works." *TechCrunch.* September 6, 2016. https://techcrunch.com/2016/09/06/ultimate-guide-to-the-news-feed/.

32. Ali, Tanveer. "How Every New York City Neighborhood Voted in the 2016 Presidential Election." *DNAinfo.* November 9, 2016. https://www.dnainfo.com

/new-york/numbers/clinton-trump-president-vice-president-every-neighbor
hood-map-election-results-voting-general-primary-nyc.

33. Gottfried, Jeffrey, and Elisa Shearer. "News Use Across Social Media Platforms 2016." Pew Research Center. May 26, 2016. http://www.journalism.org/
2016/05/26/news-use-across-social-media-platforms-2016/.

34. Briener, Andrew. "Pizzagate, explained: Everything you want to know about
the Comet Ping Pong pizzeria conspiracy theory but are too afraid to search
for on Reddit." *Salon.* December 10, 2016. http://www.salon.com/2016/12/10
/pizzagate-explained-everything-you-want-to-know-about-the-comet-ping
-pong-pizzeria-conspiracy-theory-but-are-too-afraid-to-search-for-on-reddit/.

35. Williams, Rhiannon. "Facebook: 'We cannot become arbiters of truth—it's
not our role.'" *iNews.* April 6, 2017. https://inews.co.uk/essentials/news/tech
nology/facebook-looks-choke-fake-news-cutting-off-financial-lifeline/.

36. "News Use Across Social Media Platforms 2016."

37. Pogue, David. "What Facebook Is Doing to Combat Fake News." *Scientific
American.* February 1, 2017. https://www.scientificamerican.com/article/pogue
-what-facebook-is-doing-to-combat-fake-news/.

38. Harris, Sam. *Free Will* (New York: Free Press, 2012), 8.

39. Bosker, Bianca. "The Binge Breaker." *Atlantic*, November 2016. https://www
.theatlantic.com/magazine/archive/2016/11/the-binge-breaker/501122/.

Chapter 5: Google

1. Dorfman, Jeffrey. "Religion Is Good for All of Us, Even Those Who Don't Follow One." *Forbes.* December 22, 2013. https://www.forbes.com/sites/jeffrey-
dorfman/2013/12/22/religion-is-good-for-all-of-us-even-those-who-dont
-follow-one/#797407a64d79.

2. Barber, Nigel. "Do Religious People Really Live Longer?" *Psychology Today.*
February 27, 2013. https://www.psychologytoday.com/blog/the-human-beast
/201302/do-religious-people-really-live-longer.

3. Downey, Allen B. "Religious affiliation, education, and Internet use." arXiv.
March 21, 2014. https://arxiv.org/pdf/1403.5534v1.pdf.

4. Alleyne, Richard. "Humans 'evolved' to believe in God." *Telegraph.* September 7,
2009. http://www.telegraph.co.uk/journalists/richard-alleyne/6146411/Hu
mans-evolved-to-believe-in-God.html.

5. Winseman, Albert L. "Does More Educated Really = Less Religious?" Gallup.
February 4, 2003. http://www.gallup.com/poll/7729/does-more-educated-really
-less-religious.aspx.

6. Rathi, Akshat. "New meta-analysis checks the correlation between intelligence and faith." *Ars Technica.* August 11, 2013. https://arstechnica.com/science/2013/08/new-meta-analysis-checks-the-correlation-between-intelligence-and-faith/.

7. Carey, Benedict. "Can Prayers Heal? Critics Say Studies Go Past Science's Reach." *New York Times.* October 10, 2004. http://www.nytimes.com/2004/10/10/health/can-prayers-heal-critics-say-studies-go-past-sciences-reach.html.

8. Poushter, Jacob. "2. Smartphone ownership rates skyrocket in many emerging economies, but digital divide remains." Pew Research Center. February 22, 2016. http://www.pewglobal.org/2016/02/22/smartphone-ownership-rates-skyrocket-in-many-emerging-economies-but-digital-divide-remains/.

9. "Internet Users." Internet Live Stats. http://www.internetlivestats.com/internet-users/.

10. Sharma, Rakesh. "Apple Is Most Innovative Company: PricewaterhouseCooper (AAPL)." Investopedia. November 14, 2016. http://www.investopedia.com/news/apple-most-innovative-company-pricewaterhousecooper-aapl/.

11. Strauss, Karsten. "America's Most Reputable Companies, 2016: Amazon Tops the List." *Forbes.* March 29, 2016. https://www.forbes.com/sites/karstenstrauss/2016/03/29/americas-most-reputable-companies-2016-amazon-tops-the-list/#7967310a3712.

12. Elkins, Kathleen. "Why Facebook is the best company to work for in America." *Business Insider.* April 27, 2015. http://www.businessinsider.com/facebook-is-the-best-company-to-work-for-2015-4.

13. Clark, Jack. "Google Turning Its Lucrative Web Search Over to AI Machines." Bloomberg. October 26, 2015. https://www.bloomberg.com/news/articles/2015-10-26/google-turning-its-lucrative-web-search-over-to-ai-machines.

14. Schuster, Dana. "Marissa Mayer spends money like Marie Antoinette." *New York Post.* January 2, 2016. http://nypost.com/2016/01/02/marissa-mayer-is-throwing-around-money-like-marie-antoinette/.

15. "Alphabet Announces Third Quarter 2016 Results." Alphabet Inc. October 27, 2016. https://abc.xyz/investor/news/earnings/2016/Q3_alphabet_earnings/.

16. Alphabet Inc., Form 10-K for the Period Ending December 31, 2016 (filed January 27, 2017), p. 23, from Alphabet Inc. website. https://abc.xyz/investor/pdf/20161231_alphabet_10K.pdf.

17. Yahoo! Finance. Accessed in February 2016. https://finance.yahoo.com/.

18. Godman, David. "What is Alphabet . . . in 2 minutes." CNN Money. August 11, 2015. http://money.cnn.com/2015/08/11/technology/alphabet-in-two-minutes/.

19. Basu, Tanya. "New Google Parent Company Drops 'Don't Be Evil' Motto." *Time.* October 4, 2015. http://time.com/4060575/alphabet-google-dont-be-evil/.

20. http://www.internetlivestats.com/google-search-statistics/.

21. Sullivan, Danny. "Google now handles at least 2 trillion searches per year." *Search Engine Land.* May 24, 2016. http://searchengineland.com/google-now -handles-2-999-trillion-searches-per-year-250247.

22. Segal, David. "The Dirty Little Secrets of Search." *New York Times.* February 12, 2011. http://www.nytimes.com/2011/02/13/business/13search.html.

23. Yahoo! Finance. https://finance.yahoo.com/.

24. Pope, Kyle. "Revolution at *The Washington Post.*" *Columbia Journalism Review.* Fall/Winter 2016. http://www.cjr.org/q_and_a/washington_post_bezos_am azon_revolution.php.

25. Seeyle, Katharine Q. "The Times Company Acquires About.com for $410 Million." *New York Times.* February 18, 2005. http://www.nytimes.com/2005 /02/18/business/media/the-times-company-acquires-aboutcom-for-410 -million.html.

26. Iyer, Bala, and U. Srinivasa Rangan. "Google vs. the EU Explains the Digital Economy." *Harvard Business Review.* December 12, 2016. https://hbr.org/2016 /12/google-vs-the-eu-explains-the-digital-economy.

27. Drozdiak, Natalia, and Sam Schechner. "EU Files Additional Formal Charges Against Google." *Wall Street Journal.* July 14, 2016. https://www.wsj.com/arti cles/google-set-to-face-more-eu-antitrust-charges-1468479516.

Chapter 6: Lie to Me

1. Hamilton, Alexander. *The Papers of Alexander Hamilton,* vol. X, *December 1791–January 1792.* Edited by Harold C. Syrett and Jacob E. Cooke (New York: Columbia University Press, 1966), 272.

2. Morris, Charles R. "We Were Pirates, Too." *Foreign Policy.* December 6, 2012. http://foreignpolicy.com/2012/12/06/we-were-pirates-too.

3. Gladwell, Malcolm. "Creation Myth." *New Yorker.* May 16, 2011. http://www .newyorker.com/magazine/2011/05/16/creation-myth.

4. Apple Inc. "The Computer for the Rest of Us." Commercial, 35 seconds. 2007. https://www.youtube.com/watch?v=C8jSzLAJn6k.

5. "Testimony of Marissa Mayer. Senate Committee on Commerce, Science, and Transportation. Subcommittee on Communications, Technology, and the Internet Hearing on 'The Future of Journalism.'" The Future of Journalism.

May 6, 2009. https://www.gpo.gov/fdsys/pkg/CHRG-111shrg52162/pdf/CHRG -111shrg52162.pdf.

6. Ibid.

7. Ibid.

8. Ibid.

9. Ibid.

10. Warner, Charles. "Information Wants to Be Free." *Huffington Post.* February 20, 2008. http://www.huffingtonpost.com/charles-warner/information-wants -to-be-f_b_87649.html.

11. Manson, Marshall. "Facebook Zero: Considering Life After the Demise of Organic Reach." *Social@Ogilvy, EAME.* March 6, 2014. https://social.ogilvy .com/facebook-zero-considering-life-after-the-demise-of-organic-reach.

12. Gladwell, Malcolm. "Creation Myth." *New Yorker.* May 16, 2011. http://www. newyorker.com/magazine/2011/05/16/creation-myth.

13. Alderman, Liz. "Uber's French Resistance." *New York Times.* June 3, 2015. https://www.nytimes.com/2015/06/07/magazine/ubers-french-resistance .html?_r=0.

14. Diamandis, Peter. "Uber vs. the Law (My Money's on Uber)." *Forbes.* September 8, 2014. http://www.forbes.com/sites/peterdiamandis/2014/09/08/uber-vs- the-law-my-moneys-on-uber/#50a69d201fd8.

Chapter 7: Business and the Body

1. Satell, Greg. "Peter Thiel's 4 Rules for Creating a Great Business." *Forbes.* October 3, 2014. https://www.forbes.com/sites/gregsatell/2014/10/03/peter-thiels -4-rules-for-creating-a-great-business/#52f096f754df.

2. Wohl, Jessica. "Wal-mart U.S. sales start to perk up, as do shares." Reuters. August 16, 2011. http://www.reuters.com/article/us-walmart-idUSTRE77F0KT 20110816.

3. Wilson, Emily. "Want to live to be 100?" *Guardian.* June 7, 2001. https://www .theguardian.com/education/2001/jun/07/medicalscience.healthandwell being.

4. Ibid.

5. Ibid.

6. Huggins, C. E. "Family caregivers live longer than their peers." Reuters. October 18, 2013. http://www.reuters.com/article/us-family-caregivers-idUSBRE99H12 I20131018.

7. Fisher, Maryanne L., Kerry Worth, Justin R. Garcia, and Tami Meredith. (2012). Feelings of regret following uncommitted sexual encounters in Canadian university students. *Culture, Health & Sexuality* 14: 45–57. doi: 10.1080/13691058.2011.619579.

8. "'Girls & Sex' and the Importance of Talking to Young Women About Pleasure." NPR. March 29, 2016. http://www.npr.org/sections/health-shots/2016/03/29/472211301/girls-sex-and-the-importance-of-talking-to-young-women-about-pleasure.

9. "The World's Biggest Public Companies: 2016 Ranking." *Forbes.* https://www.forbes.com/companies/estee-lauder.

10. "The World's Biggest Public Companies: 2016 Ranking." *Forbes.* https://www.forbes.com/companies/richemont.

11. "LVMH: 2016 record results." Nasdaq. January 26, 2017. https://globenewswire.com/news-release/2017/01/26/911296/0/en/LVMH-2016-record-results.html.

12. https://www.sec.gov/Archives/edgar/data/1018724/000119312517120198/d373368dex991.htm.

Chapter 8: The T Algorithm

1. Yahoo! Finance. https://finance.yahoo.com.

2. "L2 Insight Report: Big Box Black Friday 2016." L2 Inc. December 2, 2016. https://www.l2inc.com/research/big-box-black-friday-2016.

3. Sterling, Greg. "Survey: Amazon beats Google as starting point for product search." *Search Engine Land.* June 28, 2016. http://searchengineland.com/survey-amazon-beats-google-starting-point-product-search-252980.

4. "Facebook Users in the World." Internet World Stats. June 30, 2016. http://www.internetworldstats.com/facebook.htm.

5. "Facebook's average revenue per user as of 4th quarter 2016, by region (in U.S. dollars)." Statista. https://www.statista.com/statistics/251328/facebooks-average-revenue-per-user-by-region.

6. Millward, Steven. "Asia is now Facebook's biggest region." Tech in Asia. February 1, 2017. https://www.techinasia.com/facebook-asia-biggest-region-daily-active-users.

7. Thomas, Daniel. "Amazon steps up European expansion plans." *Financial Times.* January 21, 2016. https://www.ft.com/content/97acb886-c039-11e5-846f-79b0e3d20eaf.

8. "Future of Journalism and Newspapers." C-SPAN. Video, 5:38:37. May 6, 2009. https://www.c-span.org/video/?285745-1/future-journalism-newspapers& start=4290.

9. Wiblin, Robert. "What are your chances of getting elected to Congress, if you try?" *80,000 Hours.* July 2, 2015. https://80000hours.org/2015/07/what-are -your-odds-of-getting-into-congress-if-you-try.

10. Dennin, James. "Apple, Google, Microsoft, Cisco, IBM and other big tech companies top list of tax-avoiders." *Mic.* October 4, 2016. https://mic.com/arti cles/155791/apple-google-microsoft-cisco-ibm-and-other-big-tech-com panies-top-list-of-tax-avoiders#.Hx5lomyBl.

11. Bologna, Michael J. "Amazon Close to Breaking Wal-Mart Record for Subsidies." Bloomberg BNA. March 20, 2017. https://www.bna.com/amazon-close -breaking-n57982085432.

12. https://www.usnews.com/best-graduate-schools/top-engineering-schools /eng-rankings/page+2

Chapter 9: The Fifth Horseman?

1. "Alibaba passes Walmart as world's largest retailer." RT. April 6, 2016. https:// www.rt.com/business/338621-alibaba-overtakes-walmart-volume/.

2. Lim, Jason. "Alibaba Group FY2016 Revenue Jumps 33%, EBITDA Up 28%." *Forbes.* May 5, 2016. https://www.forbes.com/sites/jlim/2016/05/05/alibaba -fy2016-revenue-jumps-33-ebitda-up-28/#2b6a6d2d53b2.

3. Picker, Leslie, and Lulu Yilun Chen. "Alibaba's Banks Boost IPO Size to Record of $25 Billion." Bloomberg. September 22, 2014. https://www.bloomberg. com/news/articles/2014-09-22/alibaba-s-banks-said-to-increase-ipo-size-to -record-25-billion.

4. Alibaba Group, FY16-Q3 for the Period Ending December 31, 2016 (filed January 24, 2017), p. 10, from Alibaba Group website. http://www.alibabagroup .com/en/ir/presentations/presentation170124.pdf.

5. Alibaba Group, FY16-Q3 for the Period Ending December 31, 2016 (filed January 24, 2017), p. 2, from Alibaba Group website. http://www.alibabagroup .com/en/news/press_pdf/p170124.pdf.

6. "Alibaba's Banks Boost IPO Size to Record of $25 Billion."

7. "Alibaba Group Holding Ltd: NYSE:BABA:AMZN." Google Finance. Accessed April 12, 2017. https://www.google.com/finance?chdnp=1&chdd=1&chds=1& chdv=1&chvs=Logarithmic&chdeh=0&chfdeh=0&chdet=1467748800000&

chddm=177905&chls=IntervalBasedLine&cmpto=INDEXSP%3A.INX%
3BNASDAQ%3AAMZN&cmptdms=0%3B0&q=NYSE%3ABABA&ntsp=0&
fct=big&ei=7vl7V7G5O4iPjAL-pKiYDA.

8. Wells, Nick. "A Tale of Two Companies: Matching up Alibaba vs. Amazon."
   CNBC. May 5, 2016. http://www.cnbc.com/2016/05/05/a-tale-of-two-com
   panies-matching-up-alibaba-vs-amazon.html.

9. "The World's Most Valuable Brands." *Forbes.* May 11, 2016. https://www.forbes
   .com/powerful-brands/list/#tab:rank.

10. Einhorn, Bruce. "How China's Government Set Up Alibaba's Success."
    Bloomberg. May 7, 2014. https://www.bloomberg.com/news/articles/2014-05-07/
    how-chinas-government-set-up-alibabas-success.

11. "Alibaba's Political Risk," *Wall Street Journal.* September 19, 2014. https://www
    .wsj.com/articles/alibabas-political-risk-1411059836.

12. Cendrowski, Scott. "Investors Shrug as China's State Press Slams Alibaba for
    Fraud." *Fortune.* May 17, 2016. http://fortune.com/2016/03/17/investors-shrug
    -as-chinas-state-press-slams-alibaba-for-fraud/.

13. Gough, Neil, and Paul Mozur. "Chinese Government Takes Aim at E-Com-
    merce Giant Alibaba Over Fake Goods." *New York Times.* January 28, 2015.
    https://bits.blogs.nytimes.com/2015/01/28/chinese-government-takes-aim
    -at-e-commerce-giant-alibaba/.

14. "JACK MA: It's hard for the US to understand Alibaba." Reuters. June 3, 2016.
    http://www.businessinsider.com/r-amid-sec-probe-jack-ma-says-hard-for-us
    -to-understand-alibaba-media-2016-6.

15. DeMorro, Christopher. "How Many Awards Has Tesla Won? This Infographic
    Tells Us." Clean Technica. February 18, 2015. https://cleantechnica.com/2015
    /02/18/many-awards-tesla-won-infographic-tells-us/.

16. Cobb, Jeff. "Tesla Model S Is World's Best-Selling Plug-in Car for Second Year
    in a Row." GM-Volt. January 20, 2017. http://gm-volt.com/2017/01/27/tesla
    -model-s-is-worlds-best-selling-plug-in-car-for-second-year-in-a-row/.

17. Hull, Dana. "Tesla Says It Received More Than 325,000 Model 3 Reserva-
    tions." Bloomberg. April 7, 2016. https://www.bloomberg.com/news/articles
    /2016-04-07/tesla-says-model-3-pre-orders-surge-to-325-000-in-first-week.

18. "Tesla raises $1.46B in stock sale, at a lower price than its August 2015 sale:
    IFR." Reuters. May 20, 2016. http://www.cnbc.com/2016/05/20/tesla-raises
    -146b-in-stock-sale-at-a-lower-price-than-its-august-2015-sale-ifr.html.

19. "Tesla isn't just a car, or brand. It's actually the ultimate mission—the mother
    of all missions . . ." Tesla. December 9, 2013. https://forums.tesla.com/de_AT

/forum/forums/tesla-isnt-just-car-or-brand-its-actually-ultimate-mission-mother-all-missions.

20. L2 Inc. "Scott Galloway: Switch to Nintendo." YouTube. March 30, 2017. https://www.youtube.com/watch?v=UwMhGsKeYo4&t=3s.

21. Shontell, Alyson. "Uber is the world's largest job creator, adding about 50,000 drivers per month, says board member." *Business Insider.* March 15, 2015. http://www.businessinsider.com/uber-offering-50000-jobs-per-month-to-drivers-2015-3.

22. Uber Estimate. http://uberestimator.com/cities.

23. Nelson, Laura J. "Uber and Lyft have devastated L.A.'s taxi industry, city records show." *Los Angeles Times.* April 14, 2016. http://www.latimes.com/local/lanow/la-me-ln-uber-lyft-taxis-la-20160413-story.html.

24. Schneider, Todd W. "Taxi, Uber, and Lyft Usage in New York City." February 2017. http://toddwschneider.com/posts/taxi-uber-lyft-usage-new-york-city/.

25. "Scott Galloway: Switch to Nintendo."

26. Deamicis, Carmel. "Uber Expands Its Same-Day Delivery Service: 'It's No Longer an Experiment'." *Recode.* October 14, 2015. https://www.recode.net/2015/10/14/11619548/uber-gets-serious-about-delivery-its-no-longer-an-experiment.

27. Smith, Ben. "Uber Executive Suggests Digging Up Dirt on Journalists." BuzzFeed. November 17, 2014. https://www.buzzfeed.com/bensmith/uber-executive-suggests-digging-up-dirt-on-journalists?utm_term=.rcBNNLypG#.bhlEEWy0N.

28. Warzel, Charlie. "Sexist French Uber Promotion Pairs Riders With 'Hot Chick' Drivers." BuzzFeed. October 21, 2014. https://www.buzzfeed.com/charliewarzel/french-uber-bird-hunting-promotion-pairs-lyon-riders-with-a?utm_term=.oeNgLXer7#.boMKaOG9q.

29. Welch, Chris. "Uber will pay $20,000 fine in settlement over 'God View' tracking." *The Verge.* January 6, 2016. https://www.theverge.com/2016/1/6/10726004/uber-god-mode-settlement-fine.

30. Fowler, Susan J. "Reflecting on One Very, Very Strange Year at Uber." February 19, 2017. https://www.susanjfowler.com/blog/2017/2/19/reflecting-on-one-very-strange-year-at-uber.

31. Empson, Rip. "Black Car Competitor Accuses Uber Of DDoS-Style Attack; Uber Admits Tactics Are 'Too Aggressive.'" *TechCrunch.* January 24, 2014. https://techcrunch.com/2014/01/24/black-car-competitor-accuses-uber-of-shady-conduct-ddos-style-attack-uber-expresses-regret/.

32. "Drive with Uber." Uber. https://www.uber.com/a/drive-pp/?exp=nyc.

33. Isaac, Mike. "What You Need to Know About #DeleteUber." *New York Times.* January 31, 2017. https://www.nytimes.com/2017/01/31/business/delete-uber .html?_r=0.

34. "Our Locations." Walmart. http://corporate.walmart.com/our-story/our-lo cations.

35. Peters, Adele. "The Hidden Ecosystem of the Walmart Parking Lot." *Fast Company.* January 3, 2014. https://www.fastcompany.com/3021967/the-hid den-ecosystem-of-the-walmart-parking-lot.

36 http://www.andnowuknow.com/buyside-news/walmarts-strategy-under -marc-lore-unfolding-prices-and-costs-cut-online/jessica-donnel/53272# .WUdVw4nyvMU.

37. "Desktop Operating System Marketshare." Net Marketshare. https://www .netmarketshare.com/operating-system-market-share.aspx?qprid=10&qp customd=0.

38. "About Us." LinkedIn. https://press.linkedin.com/about-linkedin.

39. Bose, Apurva. "Numbers Don't Lie: Impressive Statistics and Figures of Link-edIn." BeBusinessed.com. February 26, 2017. http://bebusinessed.com/linked in/linkedin-statistics-figures/.

40. International Business Machines Corporation. Annual Report for the Period Ending December 31, 2016 (filed February 28, 2017), p. 42, from International Business Machines Corporation website. https://www.ibm.com/investor/fi nancials/financial-reporting.html.

Chapter 10: The Four and You

1. "Do you hear that? It might be the growing sounds of pocketbooks snapping shut and the chickens coming home . . ." AEIdeas, August 2016. http://bit.ly /2nHvdfr.

2. *Irrational Exuberance,* Robert Shiller. http://amzn.to/2o98DZE.

3. https://www.nytimes.com/2017/03/14/books/henry-lodge-dead-co-author -younger-next-year.html?_r=1.

Chapter 11: After the Horsemen

1. Yahoo! Finance. https://finance.yahoo.com/.

2. Facebook, Inc. https://newsroom.fb.com/company-info/.

3. Yahoo! Finance. https://finance.yahoo.com/.

4. "The World's Biggest Public Companies." *Forbes*. May 2016. https://www.forbes .com/global2000/list/.

5. Ibid.

6. Yahoo! Finance. https://finance.yahoo.com/.

7. "France GDP." Trading Economics. 2015. http://www.tradingeconomics.com /france/gdp.

8. Yahoo! Finance. https://finance.yahoo.com/.

9. "The World's Biggest Public Companies." *Forbes*. May 2016. https://www.forbes .com/global2000/list/.

10. Yahoo! Finance. https://finance.yahoo.com/.

11. "The World's Biggest Public Companies."

12. Facebook, Inc. https://newsroom.fb.com/company-info/.

13. "The World's Biggest Public Companies."

Index

Index